T0224691

ASTRONOMICAL SPECTROSCOPY

An Introduction to the Atomic and
Molecular Physics of Astronomical Spectra

Imperial College Press Advanced Physics Texts

Imperial College Press Advanced Physics Texts – Vol. 2

ASTRONOMICAL SPECTROSCOPY

An Introduction to the Atomic and Molecular Physics of Astronomical Spectra

JONATHAN TENNYSON

UNIVERSITY COLLEGE LONDON, UK

Imperial College Press

Published by

Imperial College Press
57 Shelton Street
Covent Garden
London WC2H 9HE

Distributed by

World Scientific Publishing Co. Pte. Ltd.
5 Toh Tuck Link, Singapore 596224
USA office: 27 Warren Street, Suite 401-402, Hackensack, NJ 07601
UK office: 57 Shelton Street, Covent Garden, London WC2H 9HE

British Library Cataloguing-in-Publication Data
A catalogue record for this book is available from the British Library.

First published 2005
Reprinted 2007

ASTRONOMICAL SPECTROSCOPY
An Introduction to the Atomic and Molecular Physics of Astronomical Spectra

ISBN 978-1-86094-513-7
ISBN 978-1-86094-529-8 (pbk)

Typeset by Stallion Press
Email: enquiries@stallionpress.com

Printed in Singapore

PREFACE

This book follows closely a lecture course I gave entitled 'Astronomical Spectroscopy' to third-year undergraduate students at University College London between 1998 and 2003. The students who attended had done a prior introductory course on Quantum Mechanics which covered the hydrogen atom but no further atomic physics or spectroscopy. A similar level of prior knowledge is assumed in the current work.

There are many people whose help have been essential for the completion of this book. First I must thank Bill Somerville who inaugurated the course Astronomical Spectroscopy and taught it for two years before me. He selflessly shared his lecture notes and other materials with me. I would like to thank Ceinwen Sanderson for turning my hand-scrawled lecture notes into LATEX, and my colleagues Tony Lynas-Gray, Bill Somerville, Peter Storey and Jeremy Yates for their extensive comments on the draft of the book. I owe a debt of gratitude to my graduate students Bob Barber and Natasha Doss who checked all the problems and found many errors. I thank all of them for the corrections; any errors that remain are all mine.

I must also thank the students who attended my Astronomical Spectroscopy course. It was great fun to teach, not least because the latest developments in astrophysics often fed straight into the lectures. Particular thanks are due to the class of 2003 who made a number of helpful comments and suggestions on the contents of the book.

A book on spectroscopy thrives on good illustrations and I have shamelessly plundered the literature and other sources for spectra to illustrate this one. I must thank Xiaowei Liu for help with digitising many of the published spectra, my student Iryna Rozum, my son Matthew, and especially David Rage for their help with the other illustrations.

I thank the journal publishers and the many authors who greeted my requests to reproduce their work with prompt enthusiasm, especially those authors who adapted figures at my request. Each journal and author is individually acknowledged in the figure captions.

Finally I must acknowledge the UCL astronomers of the past and present who have answered my many questions on astrophysics with a patience their frequent stupidity probably did not deserve. Particularly high on this list are Pete Storey and Mike Barlow, but the rest of the varied lunch crew should not be forgotten. Without you my knowledge of things astronomical would be the same as it was the day I arrived at UCL — nothing.

Jonathan Tennyson
London,
July 2004

CONTENTS

CHAPTER ONE

WHY RECORD SPECTRA OF ASTRONOMICAL OBJECTS?

'We will never know how to study by any means the chemical composition (of stars), or their mineralogical structure'

– Auguste Comte (1835)

1.1 A Historical Introduction

In the first part of the 19th century, astronomers began to make parallax measurements which revealed for the first time how distant even the closest stars are from us. Since travel to the stars was, and still is, impossible with foreseeable technology, many scientists believed that the composition and character of the stars would forever remain a mystery. This view is pithily summarised by the quote from the positivist French philosopher Auguste Comte (1798–1857) given above.

Today, the composition of stars, and indeed of the diffuse material in the large spaces in between the stars, is well known. How did this situation come about? In fact the first steps to finding the solution to the problem had been taken even before Comte began writing.

In 1814, Joseph von Fraunhofer (1787–1826) used one of the high-quality prisms he had manufactured to diffract a beam of sunlight, taken from a slit in his shutters, onto a whitewashed wall. Besides the characteristic colours of the rainbow, which had been observed in this fashion since Newton, he saw many dark lines (see Fig. 1.1). He meticulously catalogued the exact wavelength of each dark line — which are still known today as

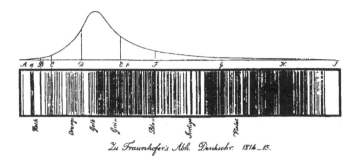

Fig. 1.1. The solar spectrum as recorded by Fraunhofer.

Fraunhofer lines — and labelled the strongest of them with letters. Many of these labels, such as the sodium D lines (see Sec. 6.4) are still used today. Fraunhofer not only recorded the first astronomical spectrum, he recorded the first-ever high-resolution spectrum. Fraunhofer's spectrum was the first to resolve discrete line transitions.

Fraunhofer did not know what caused the dark lines he observed. However he performed a similar experiment using light from the nearby red-star Betelgeuse and found that the pattern of dark lines he observed changed significantly. Fraunhofer concluded correctly that most of those features were somehow related to the composition of the object he was observing. In fact some of the lines were due to the Earth's atmosphere, the so-called telluric lines. For example, the features Fraunhofer marked A and B in his solar spectrum are actually due to molecular oxygen in our own atmosphere.

The first real step in understanding Fraunhofer's observations came in the middle of the 19th century with the experiments of Gustav Kirchhoff (1824–1887) and Robert Bunsen (1811–1899). These scientists studied the colour of the light emitted when metals were burnt in flames. They found that in certain cases the wavelength of the emitted light gave an exact match with the Fraunhofer lines. The sodium D lines, which give sodium street lights their characteristic orange colour, were one such example. These experiments demonstrated that the Fraunhofer lines were a direct consequence of the atomic composition of the Sun.

Any understanding of how these lines came about had to wait until the arrival of the 20th century with the revolution of scientific theory represented by quantum mechanics. The developments of quantum mechanics and spectroscopy have always been closely linked. As it is through the study of spectra that we have learnt of many of the riches in the Universe around us, the development of astrophysics has also been closely linked to that of spectroscopy and quantum mechanics. This book aims to give an introduction to the spectroscopy of atoms and molecules that are important for astrophysics. This book is not a text on quantum mechanics, and indeed, some basic knowledge of quantum mechanics is assumed, for it is not possible to understand or interpret spectra without some understanding of quantum mechanics.

Hearnshaw (1986) gives a fascinating historical view of the relationship between astronomy, spectroscopy and the technical developments in both fields (see further reading).

1.2 What One Can Learn from Studying Spectra

Essentially all information about astronomical objects outside the solar system comes through the study of electromagnetic radiation (light) as it reaches us. This light can contain much detailed information which is only obtained by careful analysis. Generally speaking, one can classify the information obtained by observing light according to the spectral resolution; that is the degree of sensitivity to different wavelengths, used to make the observation. One can classify such observations using the following general categories.

When one looks at the night sky with the naked eye, most astronomical bodies appear white. White light is actually light that is composed of many wavelengths which are not resolved into their different colours. Monitoring white light gives the positions of objects in the night sky. It can be used to construct maps of stars and galaxies. It can also be used to plot the movements of heavenly bodies such as comets through the night sky.

If one looks carefully at some celestial objects, such as the planets Mars and Jupiter, or stars such as Betelgeuse, one can see that these objects are tinged with a certain colour. Using instruments with low resolving power, it is possible to separate the light arriving at Earth into broad band colours. Observing colours tells us something about temperatures. For example, blue stars are hotter than red ones; objects that emit X-rays, such as the solar corona, are very hot, whereas cold objects may only emit light of very long wavelengths such as radio waves.

The most detailed astrophysical information is only obtained from high-resolution studies which involve detecting the light arriving at the earth as a function of its component wavelengths. This allows detailed spectroscopic features to be identified separately from broad band features such as colour. At the highest resolution, such studies not only yield the central wavelength of any feature, often referred to as a line, but also the shape of the feature. Such studies can yield significant extra information and this book is largely devoted to the physical basis of this information and how it can be interpreted.

To interpret an astronomical spectrum, one needs considerable knowledge of atomic and molecular physics. This knowledge usually comes from laboratory studies which provide the basic physical parameters necessary for understanding the astronomical spectrum. There is a direct relationship between these physical parameters and the astronomical information that can be obtained by observing spectra. Thus for any line observed in an astronomical spectrum, one can potentially use laboratory data to extract the following information.

The **composition** of the object being observed can be inferred by knowing which atom (or ion or molecule) produces the observed transition.

The **temperature** and other physical conditions can be deduced from assigning the actual transition being observed to precise energy levels in the atom. Transitions take place between many different states in a particular atom. Knowing which states are involved gives direct information on the degree of excitation of the system. This can be used to determine the physical conditions, such as the temperature or density of the environment local to the system.

The **abundance** of the species undergoing the transition can only be determined if the intrinsic strength of the transition being observed is known. Line strengths can be hard to determine in the laboratory. Astronomically, the strength of a transition is directly related to the number of atoms undergoing the transition under suitable conditions of optical depth (see below). Knowledge of the intensity of transitions is therefore important for determining the abundance of any species.

Motions of the species being observed relative to the earth, or indeed the whole region containing the species, lead to a shift in the wavelength of the line; this shift is known as the Doppler shift. The Doppler shift is the change in the line position from the position measured in the laboratory. This shift is given by the Doppler formula,

$$\frac{v}{c} = \frac{\Delta\lambda}{\lambda},$$
(1.1)

where v is the velocity of the source in a direction away from us, $c = 2.99792458 \times 10^8 \, \mathrm{m \cdot s^{-1}}$ is the speed of light, λ is the rest wavelength of the transition and $\Delta\lambda$ is the change in wavelength, known as the *Doppler shift*. Application of this formula requires laboratory measurement of the rest wavelength to high accuracy. Formula (1.1) is for non-relativistic motions. When an object is moving towards us, the transition is shifted to shorter wavelengths ('blue-shifted'), and when the object is moving away from us, it is shifted to longer wavelengths ('red-shifted'). It was through the monitoring of Doppler shifts of spectra of hydrogen atoms that allowed Edwin Hubble (1889–1953) to show in 1929 that our universe is uniformly expanding and so started from a single point or Big Bang.

The **pressure** or density of the environment local to the species undergoing the transitions can be monitored by observing the line profile. Such observations require particularly high resolutions. Spectral lines are broadened by collisions between species; the more frequent these collisions are, the greater the broadening. This process is called 'pressure broadening'. Lines are also broadened by the thermal motions according to the Doppler formula. Doppler broadening arises because hot species move about faster than cold ones. Both of these reveal information about the physical environment of the species being observed. However, the combined effects of pressure and temperature on the line profile can only be resolved using ultrahigh resolution observations.

Any **magnetic field** present can be monitored as certain spectral lines will be split into more than one component. Energy levels of states which possess angular momentum are split in the presence of a magnetic field. The result is that a single transition can become two or more distinct transitions. The degree of separation between these component lines depends directly on the strength of the local magnetic field. Such splittings, if observed, can therefore provide a measurement of this field.

The information obtained from such observations is the key to most astronomical knowledge. However, to interpret any astronomical spectra requires detailed information about the intrinsic properties of atomic spectra. For each atom or ion or molecule being observed, one needs to know:

(1) Its important spectral lines: these are often summarised using figures called Grotrian diagrams (see Sec. 5.4).
(2) Its energy level structure: also summarised on Grotrian diagrams.

(3) The intrinsic line strength of the transition(s) being observed.

(4) The precise rest (i.e. laboratory) wavelength of any transition observed.

Additional information is required to interpret pressure broadening of spectral lines and splitting in magnetic fields, however these topics will not be pursued in this book. Understanding and use of all this detailed spectroscopic information requires considerable knowledge of quantum mechanics.

At all wavelengths there are observed spectral lines which have yet to be identified (see for example Figs. 6.8 and 7.6). A particularly long-running current example are the diffuse interstellar bands or DIBs. This means that laboratory astrophysics, the study of astrophysical processes in the laboratory, based on either experiment or theory (or both), remains an active area of research.

Problems

Answers to problems are given at the end of the book.

1.1 While observing stars in a distant galaxy, Edwin Hubble observed discrete line emissions at 411.54 nm, 435.50 nm, 487.75 nm and 658.47 nm. There are H-atom transitions with rest wavelengths of 410.17 nm, 434.05 nm, 486.13 nm and 656.28 nm. Verify that these lines are all Doppler-shifted by the same amount. What is the speed of the distant star relative to earth? Is it moving towards us or away from us?

CHAPTER TWO

THE NATURE OF SPECTRA

'We all *know* what light is;
but it is not easy to *tell* what it is.'

– Samuel Johnson, quoted by Boswell (1776)

2.1 Transitions

All atoms have a series of discrete, quantised energy levels. In fact there
are infinitely many of them. A spectral line can be obtained when a
jump between two of these levels with different energies occurs with
light of the correct wavelength. This wavelength, λ, corresponds to the
exact energy difference, E, between the energy levels via the Planck
relationship

$$E = h\nu = \frac{hc}{\lambda}, \tag{2.1}$$

where ν is the frequency of the light and c is the speed of light. h is Planck's
constant, 6.626068×10^{-34} J $\cdot$ s, and $h\nu$ gives the energy carried by each
particle of light, known as a photon.

Each atom absorbs light at a series of characteristic wavelengths.
While individual transitions belonging to different species may acciden-
tally coincide, the whole series of lines is unique for each atom, and indeed,
every ionisation stage of each atom. It is often said that the spectrum of an
atom gives a unique fingerprint for that atom which distinguishes it from
all others. A closer analogy is that of the barcode used by supermarkets
and elsewhere to uniquely tag their products. The spectrum of an atom is

a barcode which, if we can read it, yields much detailed information about the atom.

2.2 Absorption and Emission

An atom can either absorb light, jumping to a higher-lying energy level, or emit light, dropping to a lower energy level (see Fig. 2.1). Not all transitions are equally likely; each transition is therefore characterised not only by a precise wavelength but also by a probability of that transition occurring.

Any level in an atom can absorb light, but it requires light of the correct wavelength from another source to make the atom jump to a higher energy level and for the photon to be absorbed. This is relatively easy to arrange in the laboratory but will not occur in all astronomical environments. A typical situation where an absorption spectra is observed comes from the atmospheres of stars, where the core of the star provides a continuum light source (see Fig. 2.2). This light source is approximately a black body curve with the temperature of the star. Species in the photosphere of the star are observed in absorption against this curve. This is the nature of Fraunhofer's solar spectrum (see Fig. 1.1). Absorption in the interstellar medium against a more distant star can also be observed given a suitable arrangement of the astronomical bodies.

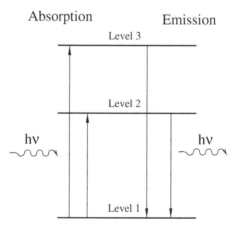

Fig. 2.1.   Emission and absorption in a schematic three-level system.

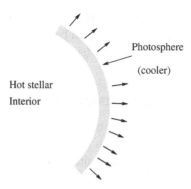

Hot stellar Interior

Photosphere

(cooler)

Fig. 2.2. The spectrum of a star: spectral lines appears as absorption against a continuous black body curve.

Emission requires that the atom involved starts in an excited state. Under such circumstances the atom can spontaneously emit a photon. The intensity of an emission spectrum from level j to level i is given by

$$N_j h \nu_{ji} A_{ji}, \qquad (2.2)$$

where N_j is the number of atoms in state j per unit volume, and $h\nu_{ji}$ gives the energy difference between the levels. A_{ji} is called the Einstein A coefficient for spontaneous emission and it gives the number of transitions per second of an atom from excited state j to state i. This is an important quantity for measuring the likelihood of a particular transition. The Einstein A coefficient, which is usually given in units of s^{-1}, will be used throughout this book to establish both transition strengths and timescales for transitions.

The intensity with which an atom in level i absorbs light to jump to level j is given by

$$N_i h \nu_{ji} B_{ij} \rho_\nu, \qquad (2.3)$$

where N_i is the number of atoms in state i, and ρ_ν is the density of radiation with frequency ν. B_{ij} is the transition probability for absorption, called the Einstein B coefficient.

Einstein proved that

$$B_{ij} = \frac{c^3}{8\pi h \nu_{ji}^3} \frac{g_j}{g_i} A_{ji}, \qquad (2.4)$$

where g_j and g_i are the statistical weights, also known as degeneracies, of states j and i respectively. This proof can be found in standard textbooks such as Bransden and Joachain (2003) (see further reading).

2.3 Other Measures of Transition Probabilities

The Einstein A_{ji} and B_{ij} coefficients can be calculated using the wavefunctions of states i and j. Although Einstein A coefficients will be used throughout this book to quantify transition probabilities, there are a number of other ways of doing this.

One often-used quantity is the oscillator strength:

$$f_{ij} = \frac{4\pi\epsilon_0 m_e}{e^2 \pi} h\nu_{ji} B_{ij},\tag{2.5}$$

where m_e is the mass of an electron and e is the charge on an electron. The factor $4\pi\epsilon_0$, where ϵ_0 is the permittivity of a vacuum, enters when the expression is given in SI units.

If g_i is the degeneracy of level i, it can be shown that

$$g_j f_{ij} = -g_i f_{ji},\tag{2.6}$$

where the convention that f is positive for absorption and negative for emission has been used. f has the advantage of being dimensionless, so it can be tabulated unambiguously. Furthermore, oscillator strengths also have interesting sum rule properties which can be used, for example, to quantify missing absorption. For more information see Woodgate (1983) in further reading. Laboratory data for atomic systems are usually presented as f-values.

2.4 Stimulated Emission

Figure 2.3 shows a third way in which an atom can undergo a transition. This is a stimulated rather than a spontaneous emission. To get a stimulated emission, a photon of the correct wavelength is required to initiate the emission of the photon. The two outgoing photons are coherent and travel in the same direction. If there are a significant number of atoms in a particular excited state, then stimulated emission can provoke a cascade of photons which can result in laser (Light Amplification by Stimulated Emission of Radiation) action.

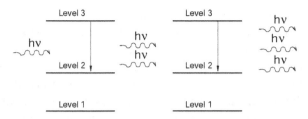

Fig. 2.3. Stimulated emission leading to laser action in a schematic three-level system.

To achieve laser action it is necessary to not only get stimulated emission but to also get population inversion: a greater population in the upper state than the lower state. The original laboratory lasers, reported by Townes and Schawlov in 1954, worked in the microwave. Maser (Microwave Amplification by Stimulated Emission of Radiation) action is also important astronomically. Specific astronomical examples will be discussed in Secs. 3.10 and 10.5.

2.5   Optical Depth

If many, many atoms of a certain species lie in the line of sight of a particular observation, then all the atoms may not be required to absorb all the light of the wavelength under study. To quantify this situation it is usual to define a quantity called the *optical depth*, τ, which is a measure of transparency of the medium. The optical depth is defined by

$$\tau = \int K \, dz, \tag{2.7}$$

where the integral runs over the path of the light being absorbed. The extinction coefficient, K, represents the product of the number density of the atoms and their opacity at the wavelength in question, where the opacity is a measure of a material's ability to absorb or block photons.

If the optical depth is large ($\tau \gg 1$), then a transition is classed as *optically thick*. Under these circumstances, it will not require all the atoms to absorb all the light at that wavelength. In this case the total absorption observed is not directly related to the number of absorbers, usually called the *column density*. Conversely, in the limit of only a few photons being absorbed, the optical depth is small ($\tau \ll 1$). In this case the line is *optically thin* and the intensity of the observed spectrum is directly proportional to the number of absorbers. Under these circumstances, the spectrum can be used to measure the column density directly.

To obtain column densities and other related information from spectra, it is important to know the optical depth of the observed transitions. Sometimes it is readily apparent that a particular transition is saturated, or optically thick (see the central feature in Fig. 3.9 for an example). However, it is often not obvious for a given observation whether a particular line is optically thin or not. In some cases a knowledge of atomic physics can be used to resolve this problem. An important example of this is discussed in Sec. 6.4.

The transport of radiation through astronomical objects is a complicated issue. It is dealt with more thoroughly by Emerson (1996) (see further reading).

2.6   Critical Density

From a spectroscopic point of view, whether a particular emission behaves as if the emitting species is in thermodynamic equilibrium with its local environment depends on something called the *critical density*. For densities above the critical density, emissions are thermal in origin, whereas below this density, each collision leads to an emission. The critical density n_c thus depends on the ratio of the timescale for decay by emission to collisional de-excitation. Specifically, the critical density for level j is given by

$$n_c = \frac{\sum_{i<j} A_{ji}}{\sum_{i \neq j} q_{ji}} , \qquad (2.8)$$

where A_{ji} is the Einstein A coefficient and the sum runs over all possible emissions, and q_{ji} is the rate for collisional de-population of level j, summed over all possible processes. This expression often simplifies to the ratio of two numbers, since for many systems and environments there is a single important route for emission and a dominant collisional de-excitation process.

The critical density is the density of collision partners, often electrons, above which the collisional de-excitation from the upper level occurs more quickly than the radiative de-excitation. In this case the emissions behave as if they are thermal and can be used as a measure of temperature. Below the critical density, each collisional excitation leads to an emission and, given knowledge of both the line strength and the collision cross section, the emissions can provide information on the density of the collision partners.

Much of the matter in the Universe is not in local thermodynamic equilibrium (LTE), as determined by the critical density. In non-LTE conditions, the concept of temperature is of limited use, although the free electrons are still usually thermalised to some temperature; such media are often characterised by the electron temperature T_e.

This definition of critical density is distinct from the cosmological critical density which determines whether the Universe is open or closed.

2.7 Wavelength or Frequency?

Measurements of spectra obtained in the laboratory are usually reported in frequency units such as Hz (which are s^{-1}) or the related wavenumber unit of cm^{-1}. Frequencies and wavenumbers are directly proportional to the energy jump of the transition. They are therefore much more amenable to physical understanding and interpretation. For this reason, most of the discussion presented below will be given in the energy/frequency/wavenumber domain.

Conversely, astronomers, at least in the infrared, visible and ultraviolet, tend to use wavelengths, typically μm, nm or Å (equal to 10^{-6}, 10^{-9} and 10^{-10} m, respectively). Spectrographs on telescopes work naturally in wavelengths. Indeed, the spectral resolution of an instrument is given by its resolving power R, as

$$R = \frac{\lambda}{\Delta\lambda},\tag{2.9}$$

where $\Delta\lambda$ is the smallest wavelength difference that can be resolved. The ratio $\frac{c}{R}$ gives the velocity resolution. Thus a resolving power of $R = 30000$ can resolve velocities in excess $10\,\mathrm{km\cdot s^{-1}}$.

The dichotomy between working in the wavelength or frequency domain means frequent application of the standard formula

$$\lambda = \frac{c}{\nu}.\tag{2.10}$$

When working in wavenumbers I find it helpful to remember that

$$\lambda(\mu m) = \frac{10000}{\nu(cm^{-1})} \, . \tag{2.11}$$

At some wavelengths, particularly at radio frequencies, it is standard practice for astronomers to present spectra as Doppler shifts from the (known) rest (that is laboratory) wavelength λ_0. This works via the Doppler formula:

$$v_r = c\frac{\Delta\lambda}{\lambda_0}, \tag{2.12}$$

where $\Delta\lambda$ represents a shift in wavelength. c is the speed of light and v_r is the velocity of the source away from us; both usually given in $km \cdot s^{-1}$. This method provides a convenient way of showing the velocity structure of the object(s) under study. Indeed, using velocity shifts means that several transitions from the same object can be shown on the same frequency scale; Figures 3.17, 3.22 and 10.19 are three of several examples given in this book. However, one note of caution must be exercised. This representation assumes that all transitions in a particular observational window have the same rest frequency λ_0. This means that transitions that are nearby in frequency but due to physically distinct processes will appear in the spectrum as velocity shifts. Examples of this are quite common: see for instance the top panel in Fig. 3.17 where the transition labelled H50β is not given relative to its rest wavelength.

2.8   The Electromagnetic Spectrum

Figure 2.4 gives an overview of the electromagnetic spectrum. The most tightly defined and smallest region in the electromagnetic spectrum is the visible region, which covers wavelengths between 4000 Å and 7000 Å. For astronomical studies, it is important to have an appreciation of the different regions of the electromagnetic spectrum for a number of reasons.

At the theoretical level, different physical processes occur at different energies, which can then be directly associated with particular regions of the spectrum. Astronomically there is also a correlation between wavelength and temperature, particularly for emission spectra. Thus, for example, very hot regions are bright at short, X-ray wavelengths, whereas cold interstellar clouds are extensively studied at long, microwave wavelengths, often referred to as radio waves.

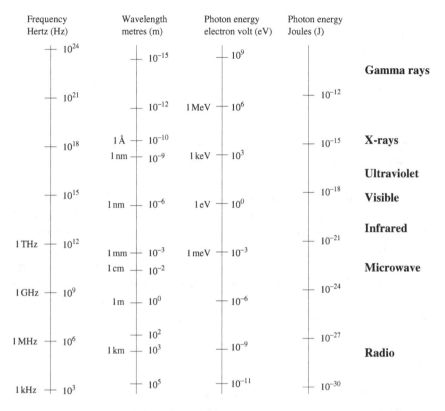

Fig. 2.4. The electromagnetic spectrum.

From an astronomical perspective there are also practical reasons for understanding the electromagnetic spectrum. All ground-based observations are necessarily made through the earth's atmosphere. The atmosphere is largely transparent at visible and radio wavelengths. Observations are possible through several infrared 'windows' in the atmosphere. These windows lie at wavelengths between regions where water vapour, and other atmospheric species absorb much of the incoming light. However in all wavelength regions there are atmospheric species which cause isolated absorptions known as 'telluric' lines. One can seek to reduce the effects of the atmosphere by the choice of good observing sites, such as the high, dry, near-equatorial sites in Hawaii and Chile, which are both home to clusters of telescopes. However there are many wavelength regions where ground-based observations are simply impossible (see Fig. 2.5).

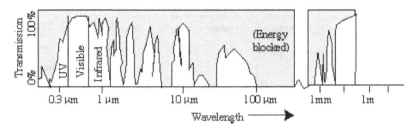

Fig. 2.5.   Absorption of light of different wavelengths in the Earth's atmosphere. Gray areas show strong telluric absorption.

The development of satellite technology over the last quarter-century means that there have now been space missions flown which study all wavelengths. Observing from space is very expensive and observing time is therefore necessarily limited. Furthermore, state-of-the-art satellite observations usually yield much poorer spectral and spatial resolutions than is obtainable on the ground because it is not practical to fly such large telescopes or such good spectrometers. Large telescopes and long integration times are also needed to observe faint objects.

Problem

2.1 The top panel of Fig. 3.17 in Chapter 3 shows two emission features labelled H40α and H50β. The velocity scale is relative to the rest frequency of H40α, which is at 99.023 GHz. Estimate the rest frequency of the H50β transition.

ATOMIC HYDROGEN

'The two most common things in the Universe are Hydrogen and Stupidity.'

– Harlan Ellison (1934)

3.1 Overview

Spectra of atomic hydrogen, H, are of paramount astronomical importance. This is because approximately 90% of atomic matter by number is hydrogen. This occurs in a variety of forms: H^+ or protons, H atoms, H_2 molecules and indeed the molecular ions H_2^+ and H_3^+. The spectral lines of hydrogen are prominent in a great variety of astronomical objects and are much studied. All aspects of hydrogen spectroscopy therefore need to be considered in detail.

The spectrum of atomic hydrogen also plays an important role in the theory of quantum mechanics. Hydrogen is the simplest atom, comprising a single electron and a proton. It is the only atom for which exact quantum mechanical solutions can be found for its energy levels and wavefunctions. These solutions will be used extensively here but will not be derived. Such derivations are a standard part of most introductory courses in quantum physics [see Rae (2002) in further reading].

3.2 The Schrödinger Equation of Hydrogen-Like Atoms

Any atom comprising a single electron orbiting a nucleus of charge Z can be described as hydrogen-like. The Hamiltonian operator for this system can be written

$$\hat{H} = \frac{-\hbar^2}{2\mu}\nabla^2 - \frac{Ze^2}{4\pi\epsilon_0 r}, \tag{3.1}$$

where $\underline{r}$ is a vector, of length r, linking the electron to the nucleus. This Hamiltonian comprises two terms. The first term, given by the Laplacian operator, is the kinetic energy operator for the electron. The second term is the potential energy term, in this case the Coulomb attraction between the electron, charge $-e$, and the nucleus, charge $+Ze$.

The Hamiltonian (3.1) contains a number of constants. It is often given in terms of atomic units, which scale the dimensions of the problem to the atomic scale. In atomic units (see Sec. 3.4 for details), the charge on an electron e, the mass of an electron m_e, $\hbar = \frac{h}{2\pi}$ and $4\pi\epsilon_0$ all equal one. This simplifies the Hamiltonian to:

$$\hat{H} = -\frac{1}{2\mu}\nabla^2 - \frac{Z}{r}. \tag{3.2}$$

In general the Schrödinger equation for a system of energy E and with wavefunction ψ is written

$$\hat{H}\psi = E\psi. \tag{3.3}$$

For the particular case of the hydrogen-like atom in atomic units, the explicit form of the Schrödinger equation is therefore

$$\left[-\frac{1}{2\mu}\nabla^2 - \frac{Z}{r} - E\right]\psi(\underline{r}) = 0. \tag{3.4}$$

3.3   Reduced Mass

The hydrogen atom is actually a two-particle problem, concerning the motion of an electron of mass m_e and a nucleus of mass M. To solve for the internal motion of this system it is necessary to rewrite the equations-of-motion of the system into one equation representing the overall translation of the whole system, with mass $M + m_e$, in space, and a second equation representing the internal motions. The Schrödinger equation (3.4) is the quantum mechanical equation representing this internal motion of the hydrogen atom once the equations governing the translational motion of the whole system through space have been separated. In this Schrödinger equation, the effective mass of the reduced system is denoted μ and is known as the reduced mass.

For a system with two particles of mass m_1 and m_2,

$$\mu = \frac{m_1 m_2}{m_1 + m_2}. \tag{3.5}$$

Clearly, if $m = m_1 = m_2$, then $\mu = \frac{m}{2}$.

For the case of the hydrogen-like atom,

$$\mu = \frac{m_e M}{M + m_e}. \tag{3.6}$$

In the limit of an infinite nuclear mass, i.e. $M = \infty$, $\mu = m_e$. In practice, M is very much bigger than m_e. For example, for H itself, M is about 1836 m_e. Under these circumstances the reduced mass is very close to m_e. However, as discussed in Secs. 3.7 and 3.11, the small shift implied by a correct treatment of the reduced mass is important for astronomical observations as it means that hydrogen and deuterium give distinct spectra that can be distinguished at moderate resolution.

3.4 Atomic Units

Atomic units provide a complete integrated unit system in the manner of SI units but with quantities scaled to the dimensions of the atom. In atomic units:

Unit of mass is the electron mass, $m_e = 1.6605402 \times 10^{-27}$ kg,

Unit of electric charge is the electron charge, $e = 1.602188 \times 10^{-19}$ C,

Unit of length, the Bohr radius $a_0 = \frac{4\pi\epsilon_0 \hbar^2}{me^2} = 5.29177249 \times 10^{-11}$ m,

Planck's constant divided by 2π, $\hbar = 1$ a.u. $= 1.05457 \times 10^{-34}$ J · s,

Similarly $4\pi\epsilon_0 = 1$.

In this unit system, the atomic unit of energy is known as the Hartree and is denoted E_h. $1E_h = 2R_\infty = 27.2113661$ eV $= 4.3597482 \times 10^{-18}$ J, where R_∞ is the Rydberg constant (see Sec. 3.7). The atomic unit of time is 2.41884×10^{-17} s and the speed of light c is 137.03599 a.u.

Atomic units are often denoted a.u. and must be carefully distinguished from the somewhat larger Astronomical Unit, AU, or indeed the less well-defined 'arbitrary units'.

3.5   Wavefunctions for Hydrogen

The Schrödinger equation (3.4) can be solved analytically by working in spherical polar coordinates, i.e. $\underline{r} = (r, \theta, \phi)$. In these coordinates the wavefunction is separable into radial and angular solutions

$$\psi(r, \theta, \phi) = R_{nl}(r)Y_{lm}(\theta, \phi). \tag{3.7}$$

The radial solutions, R_{nl}, can be expressed analytically in terms of Laguerre polynomials [see Rae (2002) in further reading]. Figure 3.1 shows both the wavefunctions and the probability distribution of the lowest few

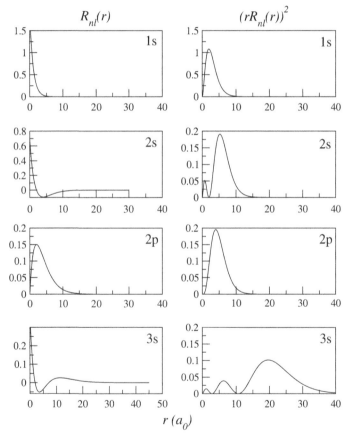

Fig. 3.1.   Wavefunctions (*left*) and probability distribution (*right*) for the radial coordinate of the hydrogen atom. (T.S. Monteiro, private communication.)

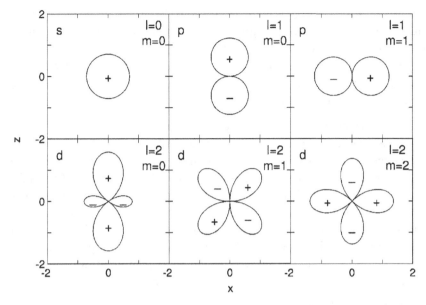

Fig. 3.2. Wavefunctions for the angular motions of the hydrogen atom which are given in terms of spherical harmonics, $Y_{l,\,m}(\theta, \phi)$. Plots are for the x–z plane except for the $l = 2$, $m = 2$ plot for which the x–y plane is shown. The plots give the absolute value of the spherical harmonic as a distance from the origin for each angle; the signs indicate the sign of the wavefunction in each region. (S.A. Morgan, private communication.)

radial solutions. An important feature of these wavefunctions is the nodes which are the points where the wavefunctions pass through zero.

The angular solutions for the hydrogen atoms, Y_{lm}, are called spherical harmonics. These functions are complex for $m \neq 0$. Figure 3.2 depicts the first few of these spherical harmonics. Readers familiar with chemistry will recognise these shapes as the s, p and d orbitals used to represent atomic structure and chemical bonds.

3.6 Energy Levels and Quantum Numbers

For bound states, the solutions of the Schrödinger equation (3.4) occur with discrete energies given by the formula

$$E_n = -\frac{\mu Z^2 e^4}{8 h^2 \epsilon_0^2} \frac{1}{n^2} = -R \frac{Z^2}{n^2}, \tag{3.8}$$

where the zero of energy is taken to be a completely separated electron and nucleus with zero kinetic energy. The constant R in Eq. (3.8) is the Rydberg constant discussed in the next section.

Each bound state of the H atom is usually characterised by a set of four quantum numbers (n, l, m, s_z). Each of these is defined below, along with a fifth quantum number s.

n is the principal quantum number. It takes the values $n = 1, 2, 3, \ldots, \infty$. n determines the energy of the atom according to Eq. (3.8).

l is the electron orbital angular momentum quantum number. The actual angular moment is given by $\hbar[l(l+1)]^{\frac{1}{2}}$. l can take the values $0, 1, 2, \ldots, n-1$. By convention, the values of l are usually designated by letters (see Table 3.1).

m is the magnetic quantum number, so called because it determines the behaviour of the energy levels in the presence of a magnetic field. $m\hbar$ is the projection of the electron orbital angular momentum, given by l, along the z-axis of the system. It can take $(2l+1)$ values $-l, -l + 1, \ldots, 0, \ldots, l-1, l$.

s is the electron spin quantum number. The electron spin angular momentum is given by $\hbar[s(s+1)]^{\frac{1}{2}}$ which, for a one-electron system, equals $\frac{\sqrt{3}}{2}\hbar$, since an electron always has spin one-half.

s_z gives the projection of the electron spin angular momentum, given by s, along the z-axis of the system. This projection is actually $\hbar s_z$. In general, s_z can take $(2s+1)$ values given by $-s, -s+1, \ldots, s-1, s$. For a one-electron system, this means s_z can take one of two values: $-\frac{1}{2}$ or $+\frac{1}{2}$.

The simplest notations for the various states of H is to denote each state by its nl quantum numbers. Thus the ground state is denoted 1s; the first excited states are 2s and 2p; the $n = 3$ states are 3s, 3p and 3d. (See Fig. 3.3.) These notations leave the m and s_z quantum numbers unspecified since these quantum numbers are really only significant for H in the presence of an external field. This means that each nl configuration is $2(2l+1)$-fold

Table 3.1. Letter designations for orbital angular momentum quantum number l.

0	1	2	3	4	5	6	7	8	...
s	p	d	f	g	h	i	k	l	...

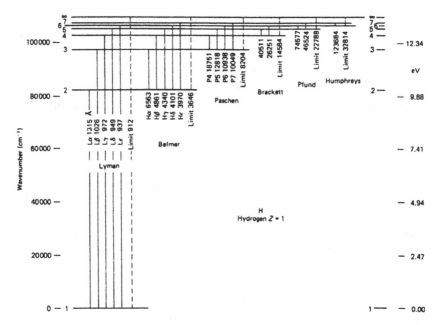

Fig. 3.3. Schematic energy levels of the hydrogen atom with various series identified. [Adapted from P.W. Merrill, *Lines of the Chemical Elements in Astronomical Spectra* (Carnegie Institute of Washington Publications, 1956).]

degenerate. Such configurations are used to build up basic atomic structure (see Sec. 4.4). More complete and complicated spectroscopic notations used to denote states of H and other atoms are discussed in Sec. 4.8.

3.7 H-Atom Discrete Spectra

The spectrum of the hydrogen atom comes from electrons jumping between different levels in the atom. Given that the energy levels depend only on the quantum number n, this means that the electronic spectrum of the H atom comes from changes in n. The wavelengths, λ, for these transitions are given by the Rydberg formula:

$$\frac{1}{\lambda} = \frac{1}{hc}(E_{n_1} - E_{n_2}) = R_\mathrm{H}\left(\frac{1}{n_1^2} - \frac{1}{n_2^2}\right), \quad n_1 < n_2. \tag{3.9}$$

This formula was constructed by the Swedish physicist Johannes Rydberg (1854–1919) to model the experimental observations of Balmer and others.

Of course, the formula agrees perfectly with the energy differences that one obtains using the quantum mechanical expression for the energy levels given in Eq. (3.8). In this expression, R is the Rydberg constant. This constant takes slightly different values for different H-like atoms since it incorporates the reduced mass μ (see Sec. 3.3). For hydrogen itself, R_H is particularly well-determined experimentally; it takes the value $109677.581\ \mathrm{cm}^{-1}$.

The Rydberg constant for a system of infinite nuclear mass is denoted by R_∞ and is related to R_H by

$$R_H = \frac{\mu}{m_e} R_\infty = \left(\frac{M_H}{M_H + m_e} \right) R_\infty, \qquad (3.10)$$

where $R_\infty = 109737.31\ \mathrm{cm}^{-1}$. Rydberg constants for other species such as deuterium, can be derived using the formula above with the appropriate reduced mass.

The spectrum of H is divided into a number of series linking different upper levels n_2 with a single lower level n_1 value. Each series is denoted according to its n_1 value and is named after its discoverer. Table 3.2 summarises the six main H-atom series.

The range of each H-atom series is given by the lowest frequency transition, between the levels n_1 and $n_2 = n_1 + 1$, and the series limit, which is the transition between n_1 and $n_2 = \infty$. As discussed in Sec. 3.8.1, the series limit is not always observable. Table 3.2 gives the spectral region in which each series is observed. As the Balmer series lies in the visible region, it is particularly easy to observe from Earth. As a result, Balmer lines have been particularly important in the study of H-atom spectra.

Table 3.2. Spectral series of the H atom. Each series comprises the transitions $n_2 - n_1$, where $n_1 < n_2 < \infty$.

				Range/cm^{-1}		
n_1	Name	Symbol	Spectral region	$n_2 = n_1 + 1$		$n_2 = \infty$
1	Lyman	Ly	ultraviolet	82257	–	109677
2	Balmer	H	visible	15237	–	27427
3	Paschen	P	infrared	5532	–	12186
4	Brackett	Br	infrared	2468	–	6855
5	Pfund	Pf	infrared	1340	–	4387
6	Humphreys	Hu	infrared	808	–	3047

Within a given series, individual transitions are labelled by Greek letters. These letters denote the change in n or Δn. In this notation:

$\Delta n = 1$ is α,
$\Delta n = 2$ is β,
$\Delta n = 3$ is γ,
$\Delta n = 4$ is δ,
$\Delta n = 5$ is ϵ.

Thus Lyα is the transition between $n_1 = 1$ and $n_2 = 2$, and Hγ is that between $n_1 = 2$ and $n_2 = 5$. Greek letters are usually only used for the most important transitions with low Δn. Transitions with high Δn are commonly labelled by the number n_2. Thus, H15 is the Balmer series transition between $n_1 = 2$ and $n_2 = 15$.

The wavelength of each individual transition can be predicted using the Rydberg formula. For example Lyα lies at

$$\frac{1}{\lambda} = R_H\left(1 - \frac{1}{4}\right) = \frac{3}{4}R_H = 82258.2\,\text{cm}^{-1}, \tag{3.11}$$

meaning that

$$\lambda = 1.21568 \times 10^{-5}\,\text{cm} = 1215.68\,\text{Å} = 121.168\,\text{nm}.$$

All hydrogen series transitions between bound states are described as bound–bound transitions. Figures 3.4, 3.5, 3.6 and 3.7 give sample H-atom spectra recorded in very different astronomical environments. Figure 3.4 shows an optical spectrum of the B-type star Θ^1 B Ori. Absorption by the Balmer series up to H14 ($n_2 = 14$) is clearly visible. Figure 3.5 shows Lyman series absorption in a shell of gas expanding about a hot, Wolf-Rayet star. Figure 3.6 shows higher members of the Paschen series recorded in absorption in a spectrum from a soft X-ray transient binary star. Figure 3.7 shows infrared emissions due to Pfund and Brackett lines recorded in the gas ejected from supernova 1987a.

The reduced mass factor means that deuterium spectral lines are shifted with respect to H-atom spectra. In principle, one can detect D and H together (see Fig. 3.8). However, the cosmic abundance of deuterium is about 2×10^{-5} that of hydrogen. This makes it difficult to measure abundance ratios directly using neighbouring transitions as the H transitions are likely to be optically thick if those of D can be observed. Figure 3.9 shows high-resolution simultaneous observations of heavily red-shifted

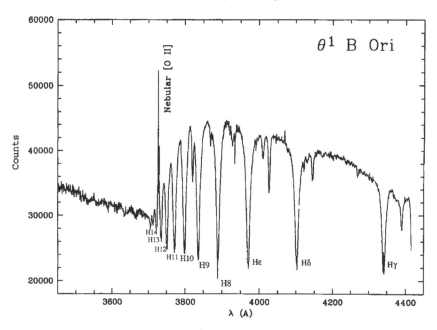

Fig. 3.4. Spectrum of B-type star Θ^1 B Ori showing Balmer series absorptions recorded at the Anglo-Australian Telescope. (X-W. Liu, private communication.)

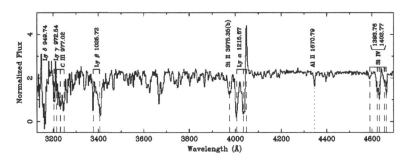

Fig. 3.5. Satellite spectrum recorded from shell nebula GRB 021004. The doublet structure in the Lyman series absorptions is caused by Doppler effects in the shell. [Adapted from N. Mirabel *et al.*, *Astrophys. J.* **595**, 935 (2003).]

H and D Lyα lines. These data need to be interpreted very carefully since H absorptions with a different Doppler shift to the main peak, such as the peak on the left of the figure, can be confused with absorptions by D. This particular spectrum was taken as part of a systematic study of

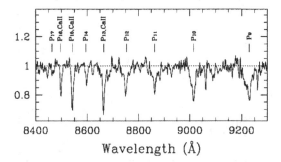

Wavelength (Å)

Fig. 3.6.   Spectrum soft X-ray transient (SRT) binary star GRO J1655-40 recorded with the Anglo-Australian Telescope, showing higher members of the Paschen series. P13, P15 and P16 are blended with the Ca II triplet transitions discussed in Sec. 6.5. [Adapted from R. Soria, K. Wu and R.W. Hunsted, *Astrophys. J.* **539**, 445 (2000).]

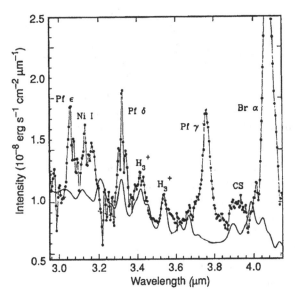

Fig. 3.7.   Infrared spectrum of supernova 1987a taken using the Anglo-Australian Telescope 192 days after the initial explosion. Strong Brackett and Pfund emissions from atomic hydrogen are clearly visible. This spectrum gave the first-ever detection of the molecule H_3^+ outside the solar system; its spectrum is modelled by the solid line. The structure in the spectrum about 3.2 μm is due to telluric effects in this region. [Analysis reported by S. Miller *et al.*, *Nature* **355**, 420 (1992).]

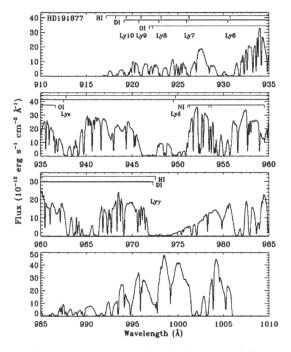

Fig. 3.8. Lyman absorption spectrum for hydrogen and deuterium recorded using the FUSE satellite along a line of sight towards B-type star HD 191877. [Adapted from C.G. Hoopes *et al.*, *Astrophys. J.* **586**, 1094 (2003).]

the absorption of light from distant quasars which has firmly established the primordal D to H ratio, one of the fundamental parameters of the Big Bang. It should be noted that since deuterium is heavier, its lines should have smaller Doppler width.

Worked Example: H-atom Hα emission occurs at 15237 cm^{-1}. At what wavenumber is the corresponding transition for the D atom?
The Rydberg formula for Hα gives:

$$15237\,\mathrm{cm}^{-1} = R_{\mathrm{H}}\left(\frac{1}{4} - \frac{1}{9}\right). \qquad (3.12)$$

The Rydberg constant for deuterium, R_{D}, is given by

$$R_{\mathrm{D}} = \frac{\mu_{\mathrm{D}}}{\mu_{\mathrm{H}}} R_{\mathrm{H}}, \qquad (3.13)$$

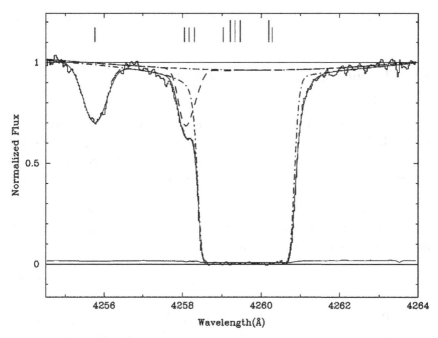

Fig. 3.9. Hubble Space Telescope spectrum of Lyα absorption against quasar QSO 1009+2956 which lies at redshift $z = 2.5$. The continuous lines give the results of models which fit several Ly series absorptions. The vertical lines give line centres; the three lines near 4258 Å are due to D; other absorption features are all due to H. [Adapted from S. Burles and D. Tytler, *Astrophys. J.* **507**, 732 (1998).]

where μ_H and μ_D are the reduced masses of the hydrogen and deuterium atom, respectively.

$$\frac{\mu_H}{\mu_D} = \frac{M_H + m_e}{M_H m_e} \frac{M_D m_e}{M_D + m_e} = \frac{(M_H + m_e)M_D}{(M_D + m_e)M_H} = 1.00027 \, , \qquad (3.14)$$

where the masses $M_H = 1836.1 \, m_e$ and $M_D = 3670.4 \, m_e$ have been used. Using these numbers gives Hα for D as 15233 cm^{-1}, compared to 15237 cm^{-1} for H.

Although the Rydberg constant is known very precisely, the wavelengths of various transitions in hydrogen do not coincide completely with the values predicted by the Rydberg formula. It should be noted that this formula is only correct within certain assumptions which neglect small effects due to relativity and the finite size of the hydrogen nucleus.

For this reason, the measured wavelengths should be used if high accuracy is desired.

3.8   H-Atom Spectra in Different Locations

3.8.1   *Balmer series*

Balmer series lines are the most studied H-atom lines since they are in the visible. Indeed, Hα emissions at 6563 Å can be clearly seen with the naked eye as the red light surrounding the Sun during a total eclipse. In fact, Hα and Hβ lines were labelled C and F respectively in Fraunhofer's solar spectrum (see Fig. 1.1).

Stellar envelopes are fairly high density environments. This means that the population of the different atomic energy levels is thermal and given by the Boltzmann distribution:

$$P_i = \frac{g_i}{Q} \exp\left(-\frac{\Delta E_i}{kT}\right), \tag{3.15}$$

where P_i is the population of the ith level given as a proportion of the atoms in level i. P_i can therefore take values between zero, meaning no atoms in level i, and one, meaning all the atoms are in level i. In Eq. (3.15), g_i is the degeneracy (statistical weight) of level i, and the energy above the ground state is ΔE_i. Q is the partition function which ensures that the sum over all populations is unity. k is Boltzmann's constant. When analysing spectra, it is useful to remember that $k = 0.695 \text{ cm}^{-1} \text{ K}^{-1}$.

However the temperature, T, varies between stars, which results in very different spectral characteristics. Since the centre of the star generates bright radiation, spectral lines are seen in absorption. The strength of these absorptions depends on the spectral type of the star.

Hydrogen Balmer lines are strongest for A0 stars which have a temperature of about 10000 K. Few other transitions are seen in these stars. Cooler stars have less population in the $n = 2$ level of H, so the Balmer line absorption diminishes. Indeed, in the coolest stars ($T < 5000$ K), molecules form and little atomic H remains. For stars significantly hotter than 10000 K, H atoms become increasing ionised and the strength of the Balmer series drops again.

This discussion illustrates a general point. Atoms lose electrons to become ions. As the environment gets hotter, the degree of ionisation increases. Since this is essentially a thermal effect, for every atom, there is a particular ionisation stage which is dominant at a particular temperature. At this temperature, the spectral lines of this ion will be at their strongest. Furthermore, successive ions of particular atoms have spectra quite unlike each other; for example H ions are protons which have no spectrum at all!

Figure 3.10 shows the temperature distribution of iron ions. The Roman numerals in this figure are used to designate ionisation stages. Thus, Fe I corresponds to neutral iron Fe; Fe II to singly ionised iron Fe^+; Fe III to Fe^{2+}, and so forth. Fe XXVI denotes Fe^{25+} which is H-like, i.e. it has one electron. Fe XXVII is a bare iron nucleus. Strictly speaking, Roman numerals label a spectrum, but they are often used to indicate an ion.

For hydrogen, it is usual to talk of H I regions where H is neutral. Often, no optical H-atom spectrum is seen from these regions. Conversely, in H II regions, the H atoms are ionised. As discussed below, H I recombination spectra are seen from H II regions.

Spectra are sensitive to pressure effects. The width of spectral lines depends on pressure, and information on the luminosity class of a star can

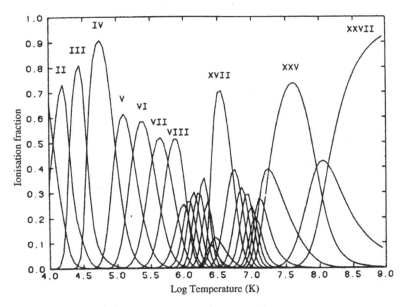

Fig. 3.10. Abundance of iron ions with different ionisation stages as a function of temperature. These results are taken from the coronal model of J.M. Schull. [Adapted from A. Dalgarno and D. Layzer (eds.), *Spectroscopy of Astrophysical Plasmas* (Cambridge University Press, 1987).]

be obtained by measuring these line-widths. If the pressure is low, higher members of the Balmer series can be seen (see Fig. 3.4). However, the complete series is never observed in stars and the spectrum shows a discontinuity known as the *Balmer jump* (see Fig. 3.11). The jump occurs at the point where continuous (bound-free) absorption switches on. The position of the Balmer discontinuity shifts to longer wavelengths at higher pressure.

To understand why, it is necessary to consider the physical size of the H atom. The approximate radius of the electron orbit in the Bohr atom, r_n, is given by the formula:

$$r_n = n^2 \frac{4\pi\epsilon_0 \hbar^2}{m_e^2 Z} = \frac{n^2}{Z} a_0 , \qquad (3.16)$$

where for H(1s), $r_1 = 1$ and $a_0 = 0.529 \times 10^{-10}$ m. At high pressure, the mean atom–atom distance will be smaller than some r_n, and higher orbits will not be found since they will be destroyed by atom–atom collisions.

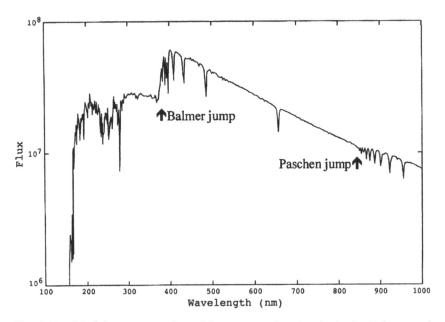

Fig. 3.11. Model spectrum of an A5-type star showing both the Balmer and Paschen discontinuities. (R.J. Sylvester, private communication.)

Worked Example: The Sun has a number density, N, of about $10^{17}\,\text{cm}^{-3}$. What is the highest H-atom n level that one would expect to find? The approximate radius of an electron orbit in the H atom is:

$$r_n = n^2 \times 0.529 \times 10^{-10}\,\text{m}.$$

Therefore the approximate volume of an H atom in state n is

$$V_n \approx \frac{4}{3}\pi r_n^3 \approx \frac{4}{3}\pi n^6 \times 1.48 \times 10^{-31}\,\text{m}^3.$$

On the surface of the Sun, $N \approx 10^{17}\,\text{cm}^{-3} = 10^{23}\,\text{m}^{-3}$. This means that each atom occupies a maximum volume of about $10^{-23}\,\text{m}^3$.

Assuming that the highest n value corresponds to the maximum volume allowed for each atom gives

$$\frac{4}{3}\pi n^6 \times 1.48 \times 10^{-31} \approx 10^{-23},$$

which gives $n \approx 16$ as the highest level that one would expect on the Sun.

3.8.2 *Lyman series*

The Lyman series is expected to be strong in absorption spectra of hot stars which have significant ultraviolet continuum. However, there are a number of technical issues that need to be considered as it is not possible to observe such ultraviolet transitions from the ground.

So far, only a few satellites with the capability of recording spectra this far into the ultraviolet have been launched. The Hubble Space Telescope, for example, does not possess this capability; its best region for spectroscopy is given by $\lambda \geq 1300\,\text{Å}$, whereas Ly$\alpha$ lies at 1216 Å. The highly successful International Ultraviolet Explorer (IUE) satellite, which operated for 19 years starting from 1978, covered Lyα but no higher Lyman lines. However, this situation changed with the launch of NASA's FUSE (Far Ultraviolet Spectroscopic Explorer) satellite in June 1999 (see Fig. 3.8). FUSE covers $\lambda = 900$–$1200\,\text{Å}$ at good resolution and sensitivity. FUSE's range thus extends below the Ly series limit at 912 Å.

Lyman absorption comes from the H-atom ground state. It is thus dominated by a strong interstellar component. See Fig. 3.9 for an example. Interstellar H atoms are all in their ground states, and are thus only sensitive to Ly wavelengths. This complicates the interpretation of any Ly emission spectrum.

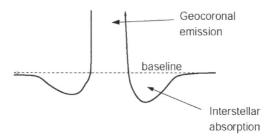

Fig. 3.12.   Schematic Lyman α emission line showing geocoronal emission and, in the wings, absorption against starlight from the interstellar medium.

Spectra taken in the locality of the earth are contaminated by strong geocoronal Lyα emission (see Fig. 3.12). Such emissions are present in IUE spectra, which were taken in geosynchronous orbit, and also some Hubble Space Telescope spectra, for which Fig. 7.7 gives an example. Figure 10.13, which was recorded with FUSE, shows the clear signature of geocoronal Lyβ emission.

Lyα can be used as a probe of the Universe at earlier epochs. Quasars emit ultraviolet light with a quasi-continuous distribution. Absorption of this light by H atoms via the Lyα transition can be monitored with different epochs being distinguished by their differing redshifts.

Figure 3.13 shows a high-resolution spectrum of a high redshift quasar. The strong and broad Lyman α emission line is observed at a wavelength of 5622 Å, indicating that the quasar is at a redshift $z = 3.625$. The spectrum of the quasar at shorter wavelengths is eaten away by a great number of narrow absorption lines. Most of these are single Lyα lines formed in gas located between the quasar and us; they therefore appear at many discrete redshifts between $z = 3.625$ and 0. These absorption lines are collectively known as the *Lyman α forest*. They are a powerful probe of physical conditions in galaxies and the intergalactic medium at early times in the evolution of the universe.

3.8.3   *Infrared lines*

The H-atom series higher than Balmer absorb and emit in the infrared. These can be seen in a number of locations. For example, Fig. 3.7 shows the spectrum of the outer expanding shell of supernova 1987a taken 192 days after the initial supernova explosion. This spectrum was used to report the

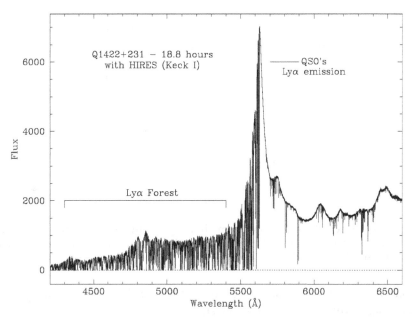

Fig. 3.13. Spectrum of the high redshift QSO Q1422+231, obtained with the High Resolution Echelle Spectrograph on the Keck telescope in Hawaii showing numerous Lyα absorptions by the intervening interstellar medium. These absorptions are known collectively as the *Lyα forest*. (M. Pettini, private communication.)

discovery of the H_3^+ molecular ion in the expanding gas, however it also shows strong Brα emission and several Pfund series lines.

3.9 H-Atom Continuum Spectra

3.9.1 *Processes*

The continuum of a proton and an electron is not quantised. Therefore an H atom in its 1s state can be ionised by any photon with $\lambda < 912$ Å, i.e. beyond the Lyman series limit.

$$H(1s) + h\nu \rightarrow H^+ + e^- . \tag{3.17}$$

This process is called *photoionisation*; it is a bound-free transition. Similarly, the Balmer continuum is observed for $\lambda < 3646$ Å (see Fig. 3.11). The probability of photoionisation is greatest when near threshold and drops away as λ decreases. See Fig. 3.14 which illustrates bound–bound and

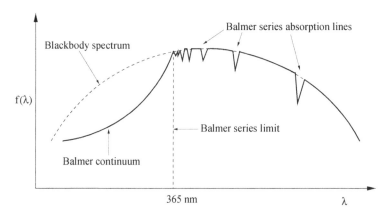

Fig. 3.14. Schematic stellar spectrum showing the hydrogen Balmer lines and continuum. There is a large departure from a black body spectrum at ultraviolet wavelengths below the Balmer series limit.

bound–free (photoionisation) spectra for Balmer lines absorbing against the black body radiation curve of a star.

The reverse process,

$$H^+ + e^- \rightarrow H(nl) + h\nu, \tag{3.18}$$

is called *radiative recombination*; it is a free–bound transition. The probability that an electron will emit a photon as it passes the proton is low, so this is a very unlikely process. However, the chances increase if the electron is travelling slowly, i.e. near threshold.

Radiative recombination is essential, if inefficient, for cooling. For example the recombination era, which started about 300000 years after the Big Bang and lasted for about 1 million years, was the period in which neutral H and He formed in the early Universe. This is the earliest epoch of our Universe for which it is theoretically possible to record spectra.

3.9.2 *H-atom emission in H II regions*

Figure 3.15 shows many higher Balmer lines recorded in Orion. These emissions come from H II regions — ionised regions around hot stars where there are strong ultraviolet fluxes and gases with temperatures of about 10000 K. Under these conditions any H atom will be rapidly photoionised.

Although the recombination of an electron and a proton to produce an H atom is an inefficient process, it still happens continually in H II

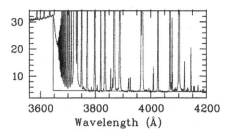

Fig. 3.15. Spectrum of the Orion nebula (M42) recorded using the ESO 1.52 m telescope in Chile. The vertical scale gives the observed flux in units of 10^{-16} ergs $\cdot$ cm$^{-2} \cdot$ s$^{-1} \cdot$ Å$^{-1} \cdot$ arcsec^{-2}. Also shown are are polynomial fits to the continuum bluewards and redwards of the Balmer jump at 3646 Å. These can be used to determine the temperature of the nebula. [Adapted from X.-W. Liu *et al.*, *Astrophys. J.* **450**, L59 (1995).]

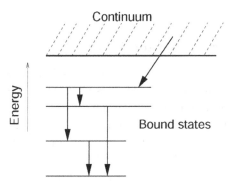

Fig. 3.16. Recombination spectra start with the emission of a continuum photon followed by a cascade of emission lines known as *recombination lines*.

regions. This produces continuum emissions and a small proportion of neutral H atoms. Typically, about 1% of the protons are in the form of neutral atomic H. Recombination can occur at different levels of the atom. If the atom is formed in an excited level, it can decay by a series of emission lines (see Fig. 3.16). Since the emission lines follow in a series, the process is sometimes referred to as a *cascade*. Different routes are possible and which is preferred depends on the relative values of the different Einstein A coefficients. The ratio of the A coefficients, which determine the relative importance of the competing pathways, are known as *branching ratios*.

H-atom lines from H II regions occur as a result of recombination from the continuum. They are called *recombination lines*.

Collisional excitation of H in lower energy levels is not important in H II regions — a ground state atom is much more likely to be photoionised. Populations of different levels are therefore determined by radiative processes and not a Boltzmann distribution.

Recombination and collisional excitation thus form competing mechanisms for driving emission spectra. In H II regions, the most common situation is that while the H-atom spectra are dominated by recombination lines, the spectra of the other atomic species present are driven by collisional excitation. This means that by careful study of both types of emission spectra, one can obtain a measure of both the density and the temperature of the nebula, as well as the strength of the radiation field which drives the ionisation process. This means that by studying spectra one can get information on both the nebula and the central star that powers it.

3.10 Radio Recombination Lines

As discussed in Sec. 3.8.1, the Balmer series is truncated in the atmospheres of stars by pressure effects. This means that the long wavelength transitions between higher levels cannot be observed. Densities in planetary nebulae are much lower, about 10^3–10^4 cm^{-3}, compared to about 10^{17} cm^{-3} at the Sun's surface region, or about 10^{19} cm^{-3} in the Earth's atmosphere at sea level. At very low densities, bound states with large n can exist. The Bohr radius for an H atom with $n = 137$ is approximately $1\,\mu$m, meaning that atoms in this state can only survive at densities significantly less than 10^{12} particles $\cdot$ cm^{-3}.

Recombination lines, transitions between neighbouring high values of n, can be detected at radio wavelengths. These transitions are labelled using a high n version of the series notation described in Sec. 3.7.

Transitions $n \leftarrow n + 1$ are called 'H$n\alpha$',

Transitions $n \leftarrow n + 2$ are called 'H$n\beta$',

and so forth. It should be noted that 'H' here stands for hydrogen and not Balmer. The transition to state n is strongest when $\Delta n = 1$, i.e. the H$n\alpha$ transition (see Fig. 3.17). Table 3.3 gives wavelengths and lifetimes for some radio-frequency transitions.

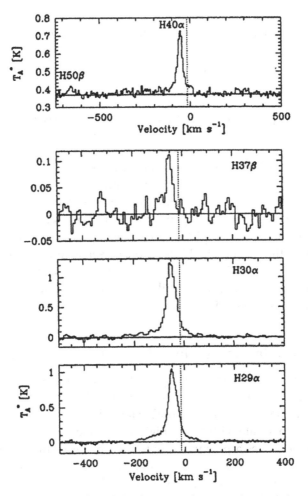

Fig. 3.17. Millimeter wavelength hydrogen radio recombination lines observed toward the very luminous, peculiar star η Carinae using the SEST telescope. Note that the intensity is given by the antennae temperature, T_A, which is conventional in radio astronomy. [Reproduced from P. Cox *et al.*, *Astron. Astrophys.* **295**, L39 (1995).]

Transitions with very high n have been observed, such as H766α, where the size of the H atom involved approaches 0.1 mm. Certain lines are observed more than others. For example, the transitions H109α and H137β both have wavelengths near 6 cm and can be observed at the same time. The H166α line is near 21 cm, where many detectors are available for observing the so-called *21 cm transition* (see Sec. 3.14).

Table 3.3.   Wavelengths, λ, and Einstein A coefficients for some typical radio frequency hydrogen atom transitions.

	H50α	H50β	H50γ	H150α
λ	5.9 mm	3.0 mm	2.1 mm	15.5 cm
A/s^{-1}	18	9.4	6.1	0.008

Radio recombination of H atoms can lead to population inversion. Population inversion is one of the conditions necessary for maser (microwave amplified stimulated emission of radiation) action (see Sec. 2.4), a process physically similar to that of a laser. In fact, H atoms can mase in dense H II regions. Such masing has been observed from MCW 349 via the $n\alpha$ lines. In this location, maser action has been observed for all transitions with $7 \leq n \leq 90$. However, molecular masers are much more common and important than those seen in atomic sources. Molecular masers will be discussed in Sec. 10.5.

3.11   Radio Recombination Lines for Other Atoms

An important feature of spectroscopy is that any atom, ion or molecule has a unique spectrum by which it can be readily identified. An exception to this general rule are the radio recombination lines. When a single electron is promoted to a state with a very high principal quantum number, n, the electron experiences a potential due to the ion core which feels like that due to a single point charge such as a proton. This is because this electron is so far away from the nucleus that the nucleus and the other electrons appear to it as if they occupy a single point.

Under these circumstances the energy levels of the outer electron satisfy the H-atom formula, Eq. (3.8), with an effective nuclear charge, Z_{eff}, equal to one. Note that a more sophisticated treatment uses quantum defect theory and replaces n^2 in Eq. (3.8) with $(n - \mu)^2$, where μ is known as the *quantum defect*. Quantum defect theory is discussed in Sec. 6.1.

Within the confines of the H-atom formula, the only difference between the high n energy levels, and hence the radio recombination spectrum, of different atoms arises from the different nuclear masses. These different masses give rise to reduced masses, and Rydberg constants, which depend on the atom in question in a fairly simple fashion. The reduced-mass factor differs because the nuclear masses, M, differ between atoms, giving a scaled Rydberg formula

$$E_n = -\frac{R_\infty}{n^2} \frac{M}{M + m_e}.$$ (3.19)

Figure 3.18 gives a sample radio recombination spectrum showing lines from hydrogen, helium and carbon. Radio recombination lines of H and He atoms have been observed from many ionised locations. Lines of heavier atoms such as C have also been detected. Traces of heavier elements, which could be S or Mg, can also be detected sometimes. However as the atoms get heavier, the frequencies of the recombination lines converge (see Table 3.4), making it very difficult to conclusively identify the species involved. Note that in Fig. 3.18 the relative strengths and profiles of He and C lines differ. This is because recombination and hence the emission lines arise from different regions. Helium is only ionised in an inner zone of the H II region which has energetic photons, so it is hot. As a result, the line is strongly Doppler broadened. Conversely, carbon is ionised in cooler regions where even the H is neutral (see Fig. 3.19), so the line is much narrower.

How does the energy balance work in the H II regions from which the recombination emissions are observed? The heating comes from incoming

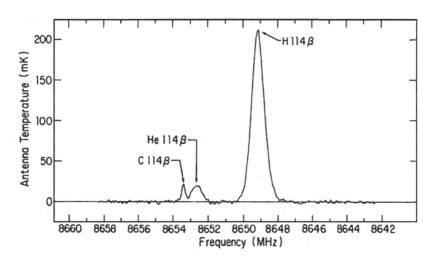

Fig. 3.18. Radio recombination lines of hydrogen, helium and carbon observed in the gaseous nebula W3. [Adapted from A. Dalgarno and D. Layzer (eds.), *Spectroscopy of Astrophysical Plasmas* (Cambridge University Press, 1987).]

Table 3.4. Frequencies, f, for the 50α radio recombination lines of hydrogen, helium, carbon and an atom with an infinite nuclear mass. The frequencies of the heavier atoms converge on this value.

	H 50α	He 50α	C 50α	$M = \infty\,50\alpha$
f/MHz	51,072	51,092	51,097	51,099

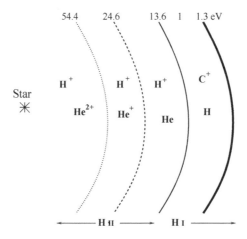

Fig. 3.19. Layers of ionised gas about a central hot star. The figures in eV give the energy required to maintain the ions indicated, and hence the maximum energy of the ultraviolet radiation, found outside each sphere.

radiation, usually from a central star, which is 'captured' by photoionisation in a bound–free transition:

$$H + h\nu \rightarrow H^+ + e^- . \qquad (3.20)$$

This is balanced by outgoing radiation, generally at longer wavelengths. This comes in the form of continuum photons formed by recombination, the reverse process to Eq. (3.20), and the resulting cascade emissions. It also comes from emissions from heavier atoms and from 'bremsstrahlung'. Bremsstrahlung, a German word meaning 'braking radiation', is generated by electrons changing velocities as they pass close to charged nuclei (see Fig. 3.20). Bremsstrahlung photons are not quantised; they can thus form a continuous distribution in frequency. Such transitions are classified

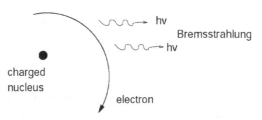

Fig. 3.20. Bremsstrahlung ('braking radiation') is emitted when a fast electron is slowed down by passing a charged nucleus.

as being free–free transitions as both the upper and lower states are not bound.

3.12 Angular Momentum Coupling in the Hydrogen Atom

One-electron atoms such as hydrogen contain several sources of angular momentum:

- Electron orbital angular momentum l;
- Electron spin angular momentum s;
- Nuclear spin angular momentum i.

The nuclear spin arises from the spin coupling of the various nucleons: protons and neutrons both have an intrinsic spin of a half. For atomic spectra it is not necessary to consider the spin of individual nucleons, just their total spin represented by the quantum number i. For many-electron atoms however, one has to consider the l and s quantum number for *each* electron in the system. This situation will be considered in Sec. 4.7.

As in classical mechanics, only the total angular momentum is a conserved quantity. It is therefore necessary to combine angular momenta together. This is best done by adding two angular momenta at a time. The order in which this is done is referred to as a coupling scheme. The choice of coupling scheme usually reflects the strength of the actual couplings: the strongest couplings are considered first.

For hydrogen, the usual coupling scheme is to combine l and s to give the total electron angular momentum j. These are added vectorially as

$$\underline{l} + \underline{s} = \underline{j}. \tag{3.21}$$

One then combines the total electron and nuclear spin angular momenta to give the final angular momentum f:

$$\underline{j} + \underline{i} = \underline{f}.\tag{3.22}$$

To do this one needs to know the rules for addition of angular momenta. This is based on the use of vector addition. Note that angular momentum is a vector as it has both a magnitude and an orientation.

In classical mechanics, adding vector $\underline{a}$ and vector $\underline{b}$ gives a vector $\underline{c}$, whose length must lie in the range

$$|a - b| \leq c \leq a + b,\tag{3.23}$$

where a, b and c are the lengths of their respective vectors. This is sometimes known as the *triangulation condition* since the lengths of the vectors must allow them to form a triangle.

In quantum mechanics a similar rule applies except that the results are quantised. The allowed values of the quantised angular momentum, c, which arise from adding a and b, span the range from the sum to the difference of a and b in steps of one:

$$c = |a - b|, |a - b| + 1, \ldots, a + b - 1, a + b.\tag{3.24}$$

For example, add the two angular momenta $l_1 = 2$ and $l_2 = 3$ together to give L. This is usually written in vector form

$$\underline{L} = \underline{l_1} + \underline{l_2},\tag{3.25}$$

and the result is:

$$L = |l_1 - l_2|, |l_1 - l_2| + 1, \ldots, l_1 + l_2 - 1, l_1 + l_2,$$
$$= 1, 2, 3, 4, 5.$$

3.13 The Fine Structure of Hydrogen

Electron spin arises as part of a relativistic treatment of quantum mechanics. Relativistic effects couple electron orbital angular momentum, l, and electron spin, s, to give the so-called fine structure in the energy levels which are split according to the value of the total electron angular momentum j.

Table 3.5. Fine structure effects in the hydrogen atom: splitting of the nl orbitals due to fine structure effect for $l = 0,1,2,3$. The resulting levels are labelled using H atom, and the more general spectroscopic notation of terms and levels (see Sec. 4.8).

Configuration	l	s	j	H atom	Term	Level
ns	0	$\frac{1}{2}$	$\frac{1}{2}$	$ns_{\frac{1}{2}}$	$n\,^2S$	$n\,^2S_{\frac{1}{2}}$
np	1	$\frac{1}{2}$	$\frac{1}{2},\frac{3}{2}$	$np_{\frac{1}{2}}, np_{\frac{3}{2}}$	$n\,^2P^o$	$n\,^2P^o_{\frac{1}{2}}, n\,^2P^o_{\frac{3}{2}}$
nd	2	$\frac{1}{2}$	$\frac{3}{2},\frac{5}{2}$	$nd_{\frac{3}{2}}, nd_{\frac{5}{2}}$	$n\,^2D$	$n\,^2D_{\frac{3}{2}}, n\,^2D_{\frac{5}{2}}$
nf	3	$\frac{1}{2}$	$\frac{5}{2},\frac{7}{2}$	$nf_{\frac{5}{2}}, nf_{\frac{7}{2}}$	$n\,^2F^o$	$n\,^2F^o_{\frac{5}{2}}, n\,^2F^o_{\frac{7}{2}}$

For hydrogen, $s = \frac{1}{2}$ so that, except for the $l = 0$ case, $j = l \pm \frac{1}{2}$ (see Table 3.5). This table labels the resulting levels with the common H-atom notation nl_j, where l is given by its letter designations, s, p, d, etc., and by spectroscopic notation for which labels of the $^{(2S+1)}L_J$ are used. A full discussion of spectroscopic notation can be found in Sec. 4.8.

Table 3.5 shows the fine structure levels of the H atom. This table shows that the states with principal quantum number $n = 2$ give rise to three fine-structure levels. In spectroscopic notation, these levels are $2\,^2S_{\frac{1}{2}}$, $2\,^2P^o_{\frac{1}{2}}$ and $2\,^2P^o_{\frac{3}{2}}$.

So far the discussion on H-atom levels has assumed that all those with the same principal quantum number, n, have the same energy. In other words, the energy does not depend on l or j. This is not correct: inclusion of relativistic (or magnetic) effects split these levels according to the total angular momentum quantum number j. This splitting, called 'fine structure', has been well-studied in the laboratory. An even more subtle effect called the *Lamb shift*, which is due to quantum electrodynamics, can also be observed. Values of these splittings for the $n = 2$ levels are given in Fig. 3.21.

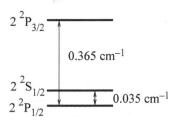

Fig. 3.21. Splitting in the $n = 2$ levels of atomic hydrogen. The larger splitting is the fine structure and the smaller one the Lamb shift.

For hydrogen, fine-structure and Lamb-shift splittings are too small to be important for most astronomical applications. The fine structure is, however, of great importance for complex atoms and will be discussed further in Chapters 4 and 6.

3.14   Hyperfine Structure in the H Atom

There is one more source of angular momentum in the H atom which has not yet been included. This is the nuclear spin, i; for H, $i = \frac{1}{2}$. Coupling i to the total electron angular momentum, j, gives the final angular momentum, f [see Eq. (3.22)]. For H this means

$$f = j \pm \frac{1}{2}. \tag{3.26}$$

The ground state of H is $1s_{\frac{1}{2}}$ or $^2S_{\frac{1}{2}}$ and has $j = \frac{1}{2}$. This means that nuclear spin coupling can split this state into two levels with $f = 0$ or 1. There is a very small, 6×10^{-5} eV, splitting between the lower $f = 0$ and higher $f = 1$ levels of H caused by magnetic effects. The $f = 0-1$ transition between these levels has a frequency of 1420.406 MHz which corresponds to a wavelength of 21 cm. The 21 cm line is probably the single most important line in astronomy. It is used to map H-atom densities throughout the ISM (see Fig. 3.22 for example).

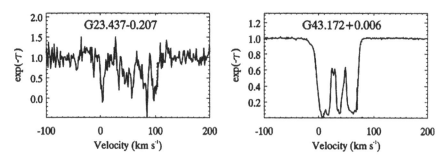

Fig. 3.22.   21-cm-line profiles observed with the Very Large Array for two galactic lines of sight recorded as part of a study which constructed a face-on galactic map of the H II region complexes. The vertical axis of the spectra represents the observed line-to-continuum intensity, which is equivalent to $\exp(-\tau_{\text{line}})$. [Adapted from M.A. Kolpak *et al.*, *Astrophys. J.* **582**, 756 (2003).]

The 21 cm line is a powerful tool because:

(1) Displacement of the line gives the line-of-sight velocity;
(2) Its intensity gives the number of atoms. Note that the line is very weak, its Einstein A coefficient is 2.9×10^{-15} s^{-1} which corresponds to a lifetime of 10 million years, so it is always optically thin;
(3) The line profile can be used to deduce the temperature of the gas. Thus, for example, Fig. 3.22 suggests that location G23.437–0.207 is much cooler that G43.172+0.006, where the lines appear to be significantly Doppler broadened.
(4) The Zeeman splitting of the transition can be used to measure the magnetic fields.

Note that the line is seen in both emission and absorption and its analysis can be complicated by the presence of several clouds along any particular line of sight.

3.15   Allowed Transitions

For hydrogen, transitions which correspond to any change in the principal quantum number, Δn, are allowed. However transitions are not observed between all states of the H atom or indeed complex atoms. Transitions are governed by selection rules which determine whether they can occur. The selection rules can be derived rigorously using the rules of quantum mechanics [see Schiff (1969) in further reading], but will simply be stated here.

Strong transitions are driven by electric dipoles. Weaker transitions, driven by both electric quadrupoles and magnetic dipoles, are observed astronomically (see Sec. 5.2), but are not important for hydrogen. Rigorous quantum mechanical analysis shows that for electric dipole transitions in hydrogen:

Δn any;
$\Delta l = \pm 1$;
$\Delta s = 0$, (for H this is always satisfied as $s = \frac{1}{2}$ for all states);
$\Delta j = 0, \pm 1$;
$\Delta m_j = 0, \pm 1$.

The last selection rule on m_j is only important in a magnetic field.

Thus if one considers the Hα transition which corresponds to $n = 2-3$, the $\Delta l = \pm 1$ condition shows that not all transitions between sub-levels occur. Specifically, the transitions 2s–3p, 2p–3s and 2p–3d are allowed whereas the transitions 2s–3s, 2p–3p and 2s–3d are not allowed.

If fine structure effects are considered, then the selection rules can give further constraints. Considering only the Hα transitions designated allowed above, the selection rule on j shows that

$$
\begin{aligned}
2\mathrm{s}_{\frac{1}{2}} - 3\mathrm{p}_{\frac{1}{2}} \quad &\text{is allowed;} \\
- 3\mathrm{p}_{\frac{3}{2}} \quad &\text{is allowed;} \\
2\mathrm{p}_{\frac{1}{2}} - 3\mathrm{d}_{\frac{5}{2}} \quad &\text{is not allowed;} \\
- 3\mathrm{s}_{\frac{1}{2}} \quad &\text{is allowed;} \\
- 3\mathrm{d}_{\frac{3}{2}} \quad &\text{is allowed;} \\
2\mathrm{p}_{\frac{3}{2}} - 3\mathrm{s}_{\frac{1}{2}} \quad &\text{is allowed;} \\
- 3\mathrm{d}_{\frac{3}{2}} \quad &\text{is allowed;} \\
- 3\mathrm{d}_{\frac{5}{2}} \quad &\text{is allowed .}
\end{aligned}
$$

3.16   Hydrogen in Nebulae

Hydrogen atom emissions in H II regions and planetary nebulae are very similar but the latter are generally brighter, which means that more weak line emissions can be observed. In particular, lines belonging to the Balmer series are often seen strongly in emission. Indeed, the characteristic red colour seen in many nebulae comes from Hα.

Balmer or other spectral series are obtained from excited atoms spontaneously emitting photons. Every excited state has a half-life τ, similar to that encountered in radioactive decay, which is related to the strength of emission. Thus excited states which decay only by weak line emission are long-lived and those which decay via strong transitions are short-lived. However, most excited states can emit to more than one other state.

The lifetime of excited state i is given by

$$
\tau_i = \left(\sum_j A_{ij} \right)^{-1} , \tag{3.27}
$$

where A_{ij} is the Einstein A coefficient (see Sec. 2.2).

Lifetimes for allowed atomic transitions are short, perhaps a few times 10^{-9} s. Table 3.6 gives some examples for the H atom. A glaring exception in Table 3.6 is the lifetime of the 2s level of H. This state has a lifetime of

Table 3.6. Lifetimes, τ, for decay by spontaneous emission for low-lying excited states of the hydrogen atom.

Level	2s	2p	3s	3p	3d
τ/s	0.14	1.6×10^{-9}	1.6×10^{-7}	5.4×10^{-9}	2.3×10^{-7}

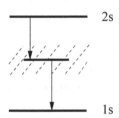

Fig. 3.23. Decay of the metastable 2s state of hydrogen giving two continuum photons.

~ 0.14 s, i.e. it lives 10^8 times longer than the 2p state. This is because the transition 2s $\rightarrow$ 1s is strongly forbidden. The 2s state is metastable which means that on the atomic scale, it is long-lived.

So how does the 2s state decay? By the process of two-photon emission, which is an inefficient process and in this case has an Einstein A coefficient of $7\,\text{s}^{-1}$ which can be compared to $A(2\text{p} \rightarrow 1\text{s}) = 6.3 \times 10^8\,\text{s}^{-1}$. The combined energy of the photons emitted must correspond to the energy difference $E(2\text{s}) - E(1\text{s})$ but the photons themselves can take any energy within this constraint (see Fig. 3.23). The photons thus appear as continuous emission radiation. Indeed the two-photon decay of the H 2s state is responsible for approximately one half the continuum emission observed from H II regions.

Problems

3.1 Give an expression for the energy levels of the hydrogen atom in terms of the Rydberg constant R_H. Assuming a value $R_H = 109677.58\,\text{cm}^{-1}$, derive a wavenumber for the Lyα transition of atomic hydrogen in cm^{-1}. Explain why the Rydberg constant, $R_\infty = 109737.31\,\text{cm}^{-1}$, is more appropriate than R_H for a heavy one-electron atom. Hence obtain an estimate for the wavenumber of the Lyα transition of hydrogen-like iron, Fe^{25+}. From what astronomical environments would such transitions occur and how might they be observed?

3.2 A proton and an electron recombine to form atomic hydrogen in its 4p state. At what wavelengths will recombination lines be observed? Label each wavelength by its standard series notation. How would the observed emissions differ if the atoms had recombined to the 4s level?

3.3 Hydrogen Hα has a rest wavelength of 6564.71 Å. At what rest wavelength would you expect to observe deuterium Hα? What telescope resolution $(R = \frac{\lambda}{\Delta\lambda})$ would be required to resolve this difference? How does this compare to the resolution required to resolve H and D Lyα transitions? Given a high enough resolution telescope, what other problems do you anticipate in obtaining the $\frac{D}{H}$ abundance by observing a single spectral line such as Hα? The mass of H is 1836.1 and D is 3670.4 in atomic units. In these units the mass of an electron is one.

3.4 A typical star has a number density, N, of about 10^{16} cm^{-3}, while in an H II region the number density is more typically 10^4 cm^{-3}. In each case what is the highest level of atomic hydrogen that is likely to be occupied? State any assumptions made in obtaining this answer. The partition function for atomic hydrogen can be written as:

$$z = \sum_{n=1}^{\infty} 2n^2 \exp\left(-\frac{R_H}{n^2 kT}\right).$$

This series cannot be summed as it diverges. Can you suggest why, in practice, the partition function for the H atom is finite?

3.5 Use the Rydberg formula to obtain the wavelength of the 80α radio line for an atom of infinite mass. Hence, taking the mass of the hydrogen nucleus to be 1836.1 electron masses, obtain the frequency of the 80α transition of atomic hydrogen. What resolving power would be required to separate the two transitions?

CHAPTER FOUR

COMPLEX ATOMS

'The truth is rarely pure and never simple.'

– Oscar Wilde, *The Importance of Being Earnest* (1895)

4.1 General Considerations

Consider an atom with N electrons and nuclear charge (atomic number) Z. It is straightforward to write down the non-relativistic Schrödinger equation for this system:

$$\left[\sum_{i=1}^{N} \left(-\frac{\hbar^2}{2m_e} \nabla_i^2 - \frac{Ze^2}{4\pi\epsilon_0 r_i} \right) + \sum_{i=1}^{N-1} \sum_{j=i+1}^{N} \frac{e^2}{4\pi\epsilon_0 |\underline{r}_i - \underline{r}_j|} - E \right]$$
$$\times \psi(\underline{r}_1, \underline{r}_2, \ldots, \underline{r}_N) = 0, \tag{4.1}$$

where $\underline{r}_i$ is the coordinate of the ith electron, with its origin at the nucleus.

The first sum in Eq. (4.1) contains a kinetic energy operator for the motion of each electron and the Coulomb attraction between that electron and the nucleus. The second summation contains the electron–electron Coulomb repulsion term. The Coulomb repulsion between pairs of electrons means the above equation is not analytically soluble, even for the simplest case, the helium atom, for which $N = 2$. This is how all atoms with two or more electrons justify their label 'complex'. To make progress on understanding these systems it is therefore necessary to introduce approximations.

4.2   Central Field Model

The easiest simplification of the complex atom problem is to try and regain a single particle problem, which is similar in spirit to the hydrogen atom problem and for which relatively easy solutions can be found. In general such solutions depend on allowing the electrons to move in a potential which does not depend on their angular position about the nucleus. Such potentials, which include the hydrogen atom potential, generate a central field since the force acting on each electron only depends on its distance from the nucleus at the centre.

Let us assume that each electron moves in its own angle-independent (central) potential given by $V_i(r_i)$ for electron i. Within this model the Schrödinger equation (4.1) can be separated into N single electron equations. This gives a simplified Schrödinger equation for the motion of the ith electron:

$$\left[-\frac{\hbar^2}{2m_e} \nabla_i^2 + V_i(r_i) \right] \phi_i(\underline{r}_i) = E_i \phi_i(\underline{r}_i) . \tag{4.2}$$

Using this model, the total energy of the system is given by the sum of single electron energies

$$E = \sum_i E_i . \tag{4.3}$$

The solutions of Eq. (4.2), $\phi_i(\underline{r}_i)$, are known as orbitals. Since much of the discussion on atomic structure is done in terms of such orbitals, it is important to emphasise that the exact solution to complex atom problems cannot be written as products of orbitals and that orbitals *only* exist with a central field or independent particle model. For this reason, this model is also often known as the *orbital approximation*.

Within the orbital approximation, each atomic orbital can be written as the product of a radial and an angular function, similar to hydrogen [see Eq. (3.7)].

$$\phi_i(\underline{r}_1) = R_{n_i l_i}(r_i) Y_{l_i m_i}(\theta_i, \phi_i) . \tag{4.4}$$

The angular part of each wavefunction is independent of the other electrons and is therefore simply a spherical harmonic $Y_{l_i m_i}(\theta_i, \phi_i)$.

So far no attempt has been made to specify the form of the central potential, $V_i(r_i)$. The standard choice is to consider the motion of each electron in the average field of all the other electrons. With this choice

$$V_i(r_i) = \frac{-Ze^2}{4\pi\epsilon_0 r_i} + \sum_{j\neq i}\left\langle \frac{e^2}{4\pi\epsilon_0 |\underline{r}_i - \underline{r}_j|} \right\rangle, \tag{4.5}$$

where the notation $\langle\ldots\rangle$ represents an average. With this form of the central potential, the radial part, $R_{n_i l_i}(r_i)$, of orbital (4.4) depends on all the other electrons. It therefore cannot be obtained in analytic form and has to be evaluated numerically using computers. As V also depends on the orbitals of the other electrons, the usual method of solution is iterative. An initial guess for the orbitals is used to generate the potential, solving Eq. (4.2) gives improved orbitals which then give a new potential, and so on, until self-consistency is reached. See Bransden and Joachain (2003) in further reading for a detailed discussion on the self-consistent field (SCF) problem.

It is standard to use the hydrogen atom orbital labels, n, l and m, to label the orbitals of other atoms. The angular behaviour, given by l and m, is indeed the same, but this is not so for the radial functions. In this case, orbitals designated by a certain n, l and m do not have the same radial structure as those given in Fig. 3.1 for hydrogen-like atoms. However they do have the same number of nodes.

Within the orbital approximation presented above, the total wavefunction for the system would be written

$$\psi(\underline{r}_1, \underline{r}_2, \ldots, \underline{r}_N) = \phi_1(\underline{r}_1)\phi_2(\underline{r}_2), \ldots, \phi_N(\underline{r}_N). \tag{4.6}$$

However this expression ignores the fact that one cannot distinguish between electron i and electron j. Before considering the indistinguishability of electrons, it is necessary to explicitly consider the role of electron spin.

To consider spin it is necessary to generalise the definition of an orbital to that of a spin-orbital:

$$\Phi_i(j) = \phi_i(\underline{r}_j, \sigma_j), \tag{4.7}$$

where $\Phi_i(j)$ means that spin-orbital i is a function of the space-spin coordinates of electron j. In Eq. (4.7), σ_j is a spin coordinate variable; it will not be necessary to explicitly define these. The four dimensions of the space-spin variable of a single particle are similar to the four vectors used to represent relativistic behaviour of a single body in classical systems.

The space-spin wavefunction for the total system can now be written:

$$\Psi(1, 2, \ldots, N) = \Phi_1(1)\Phi_2(2)\cdots\Phi_N(N). \qquad (4.8)$$

For non-relativistic treatments, such as that given by Eq. (4.1), the energy given by this wavefunction is the same as that given by ψ in Eq. (4.6).

4.3 Indistinguishable Particles

Consider a system with two identical particles. These particles can be any microscopic particles such as electrons, protons, neutrons, and so forth. If the wavefunction of these particles is $\Psi(1, 2)$ and the particles are indistinguishable, then what property must this wavefunction have?

Of course it is not the wavefunction but the probability distribution, $|\Psi|^2$, which is physically observable. If the particles are truly indistinguishable, this distribution cannot be altered by interchanging the particles. This means that

$$|\Psi(1, 2)|^2 = |\Psi(2, 1)|^2, \qquad (4.9)$$

i.e. the probability distribution is unaltered by the interchange of particle 1 and particle 2. Equation (4.9) has two possible solutions. There is the symmetric solution

$$\Psi(1, 2) = +\Psi(2, 1) \qquad (4.10)$$

or the antisymmetric solution

$$\Psi(1, 2) = -\Psi(2, 1). \qquad (4.11)$$

To explain the observed electronic structure of atoms, the German physicist Wolfgang Pauli (1900–1958) postulated that:

> Wavefunctions are antisymmetric with respect to interchange of identical Fermions.

Fermions are any particles with half-integer spin such as electrons, protons or neutrons. This statement is known as the Pauli Principle and should not be confused with the less general Pauli exclusion principle discussed below. The Pauli Principle means that for any many-electron system the wavefunction must satisfy Eq. (4.11) for each pair of electrons.

Within the orbital approximation, a two-electron wavefunction which obeys the Pauli Principle can be written

$$\Psi(1, 2) = \frac{1}{\sqrt{2}}[\Phi_a(1)\Phi_b(2) - \Phi_a(2)\Phi_b(1)] = -\Psi(2, 1), \qquad (4.12)$$

where the factor of $\frac{1}{\sqrt{2}}$ is included to keep the wavefunction normalised.

The Pauli exclusion principle arises naturally from this expression for a two-electron wavefunction. If the two spin-orbitals, Φ_a and Φ_b, are the same, i.e. $\Phi_a = \Phi_b$, then the total wavefunction, $\Psi(1, 2)$, is zero. This solution is not allowed as it cannot be normalised. Hence solutions which have the two particles occupying the same spin-orbital are inadmissible or excluded. The Pauli exclusion principle is often summarised as:

No two electrons can occupy the same spin-orbital.

This exclusion is of course the key to atomic structure and accounts naturally for the shell structures of atoms, and indeed nuclei. It also provides the degeneracy pressure which holds up the gravitational collapse of white dwarfs and neutron stars.

4.4 Electron Configurations

For a hydrogen-like atom, the energy of the individual orbitals is determined only by principal quantum number n. The energy ordering is

$$E(1s) < E(2s) = E(2p) < E(3s) = E(3p) = E(3d) < E(4s)\cdots$$

For complex atoms the situation is not so simple. The degeneracy on the orbital angular momentum quantum number l is lifted. This is because electrons in low l orbits 'penetrate', i.e. get inside orbitals with lower n-values. Penetration by the low l electrons means that they spend some of their time nearer the nucleus experiencing an enhanced Coulomb attraction. This lowers their energy relative to higher l orbitals which penetrate less or not at all. Figure 3.1 illustrates the greater penetration of s orbitals. As a result of this effect, the orbitals of complex atoms follow a revised energy ordering:

$$E(1s) < E(2s) < E(2p) < E(3s) < E(3p) < E(3d) \simeq E(4s)\cdots$$

Following the Pauli exclusion principle, each orbital labelled nl actually consists of orbitals with $2l + 1$ different m values, each with two possible values of s_z. Thus the nl orbital can hold a maximum $2(2l + 1)$ electrons.

As an example, a p orbital, for which $l = 1$, can hold up to six electrons, since the following combinations of magnetic and spin quantum numbers are possible:

$$\left(m = +1, s_z = +\frac{1}{2} \right); \ \left(m = +1, s_z = -\frac{1}{2} \right); \ \left(m = 0, s_z = +\frac{1}{2} \right);$$

$$\left(m = 0, s_z = -\frac{1}{2} \right); \ \left(m = -1, s_z = +\frac{1}{2} \right); \ \left(m = -1, s_z = -\frac{1}{2} \right).$$

An atomic configuration is given by distributing electrons amongst the orbitals. The lowest energy or ground state configuration involves filling the atomic orbitals in energy order from the lowest energy orbitals upwards.

For example, carbon has six electrons which give the following ground state configuration:

$$1s^2 2s^2 2p^2,$$

where the superscript on the orbital gives the number of electrons in that orbital. No superscript is often used to denote a single electron. In carbon, the 2p orbital contains only two electrons, which means that it is not full. Partially-filled shells usually give rise to several states with the same distribution of electrons between orbitals or configuration. This 'open shell' structure introduces complications which will be discussed further in Sec. 4.10.

The neon atom has ten electrons and its ground state configuration is

$$1s^2 2s^2 2p^6.$$

All the occupied orbitals are full and this is known as a closed shell configuration. A closed shell or sub-shell (such as $2p^6$) makes no contribution to the total orbital or spin angular momentum i.e. L or S. This property greatly simplifies the calculation of total angular momenta for complex atoms.

Atomic ions which have the same number of electrons form what are called *isoelectronic series*. These ions have the same ground state configuration. Thus, for example, all ions with ten electrons are described as neon-like. Hence Fe XVII (or Fe^{16+}) is called *neon-like iron*.

4.5   The Periodic Table

The periodic structure of the elements was originally proposed by the
Russian chemist Dmitry Ivanovich Mendeleyev (1834–1907) on the basis
of careful study of the chemical properties of each element. The structure
of the periodic table can be understood in terms of the configurations of
the individual atoms.

Table 4.1 gives the configurations of the first twenty elements in the
periodic table. In this table, closed shells have be designated using the
standard shell notation which labels the $n = 1$ orbitals as K, $n = 2$ as L,
$n = 3$ as M, $n = 4$ as N, and so forth. Note that for potassium and calcium
the M shell is not full since the 3d orbital is empty. The elements immedi-
ately after calcium, known as the transition metals, begin to progressively

Table 4.1.   Atomic configurations of the first 20 elements
in the periodic table. Z is the atomic number which corresponds to the charge on the nucleus.

Atom		Z	Configuration			
hydrogen	H	1	1s			
helium	He	2	$1s^2$			
lithium	Li	3	K	2s		
beryllium	Be	4	K	$2s^2$		
boron	B	5	K	$2s^22p$		
carbon	C	6	K	$2s^22p^2$		
nitrogen	N	7	K	$2s^22p^3$		
oxygen	O	8	K	$2s^22p^4$		
fluorine	F	9	K	$2s^22p^5$		
neon	Ne	10	K	$2s^22p^6$		
sodium	Na	11	K	L	3s	
magnesium	Mg	12	K	L	$3s^2$	
aluminium	Al	13	K	L	$3s^23p$	
silicon	Si	14	K	L	$3s^23p^2$	
phosphorus	P	15	K	L	$3s^23p^3$	
sulphur	S	16	K	L	$3s^23p^4$	
chlorine	Cl	17	K	L	$3s^23p^5$	
argon	Ar	18	K	L	$3s^23p^6$	
potassium	K	19	K	L	$3s^23p^6$	4s
calcium	Ca	20	K	L	$3s^23p^6$	$4s^2$

fill up the 3d orbital. These elements, which include the astronomically important iron, have particularly complicated electronic structures and associated optical properties.

Atoms with the same electron configuration outside a closed shell share similar chemical and optical properties. Indeed it was the observation of patterns in chemical behaviour which led to Mendeleev's original proposal of the periodic structure of the elements in 1871.

For example, lithium, sodium and potassium all have a single electron outside a closed shell. These atoms, known as the alkali metals, all have very similar optical properties. The spectra of alkali metals will be discussed in detail in Chapter 6.

Electronically-excited states of atoms usually arise when one of the outermost electrons jumps to a higher orbital. These excited states can be written as configurations. Such excited states for helium might include 1s2s, 1s2p or 1s3s. It should be noted that each of the configurations actually gives rise to more than one excited state.

States with two electrons simultaneously excited are possible but are less important. For many systems, all of these states are unstable. They have sufficient energy to autoionise by spontaneously ejecting an electron. For example, the lowest two-electron excited state of helium has the configuration $2s^2$. This state spontaneously decays to the 1s ground state of the He^+ ion and a free electron.

4.6   Ions

A neutral atom has the same number of electrons as the charge on the nucleus in atomic units. If an atom is not charge neutral it is called an ion. Under these circumstances the number of electrons, N, is not equal to the nuclear charge, Z, and the system carries a net charge of $Z - N$.

Positive ions have $N < Z$. Ions with the same number of electrons, N, and different nuclear charge, Z, are called *isoelectronic* (see Fig. 4.1). Isoelectronic ions have similar structure and spectra, and are often referred to using the neutral atom at the head of the series, such as the example neon-like iron for Fe^{16+} discussed above. The spectrum of members of the sequence is modified by the different effective charge Z_{eff}. For example, a single outer electron may feel an effective charge of $Z_{eff} = Z - N + 1$. The energies of individual levels, and hence the transition frequencies between the levels, shift in proportion to Z_{eff}^2.

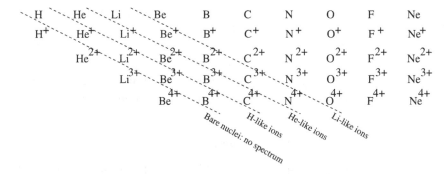

Fig. 4.1. Isoelectronic series for low Z atoms.

In the one-electron case, the 'hydrogen-like' ion transition frequencies scale closely with Z^2. For complex ions there are considerable differences in detail between spectra of ions belonging to the same isoelectronic sequence.

Negative ions have $N > Z$, although in practice only $N = Z + 1$ is important. Even then not all atoms can bind an extra electron to form a stable negative ion. Hydrogen, carbon and oxygen can bind an electron while helium and nitrogen cannot. Furthermore, most negative ions have only one stable level, and so possess no line ('bound–bound') spectrum. This means that the only possible transitions are continous bound-free absorption, also known as *photoionisation*.

In general, negative ions are much less astronomically important than positive ones. However the hydrogen anion, H^-, is a major source of opacity in cool stars including our Sun. The second electron is only bound by 0.75 eV, in contrast to the first electron which is bound by 13.6 eV. It can therefore be photoionised by all light with wavelengths less than 1.6 μm, which lies in the infrared.

4.7 Angular Momentum in Complex Atoms

Complex atoms contain more than one electron and thus have several sources of angular momentum. In particular, the ith electron has orbital angular momentum l_i, and spin angular momentum s_i, which equals $\frac{1}{2}$. There is only one conserved angular momentum in the atom. Ignoring nuclear spin effects, this angular momentum is the total (spin plus orbital) angular momentum for all electrons, J. Note that above and elsewhere the convention is followed that single electron angular momenta are represented by lower case letters, l, s, j, etc., and many electron angular momenta are represented by upper case letters, L, S, J, etc.

There are two coupling schemes or ways of summing the individual electron angular momenta to give the total angular momentum.

4.7.1 *L–S or Russell–Saunders coupling*

In Russell–Saunders coupling, the orbital and spin angular momenta of the electrons are added separately to give the total orbital angular momentum, L,

$$\underline{L} = \sum_i \underline{l}_i, \tag{4.13}$$

and the total electron spin angular momentum, S,

$$\underline{S} = \sum_i \underline{s}_i. \tag{4.14}$$

These are then added to give J

$$\underline{J} = \underline{L} + \underline{S}. \tag{4.15}$$

It is useful to remember that, as a result of the Pauli Principle, closed shells and sub-shells, such as $1s^2$ or $2p^6$, have both $L = 0$ and $S = 0$. This means that it is only necessary to consider 'active' electrons, those in open or partially-filled shells. In most cases this means only one or two electrons. When more than two angular momenta need to be added together, they should be added in pairs. The result is independent of the order in which the addition is performed.

Worked Example: Consider O III with the configuration: $1s^2 2s^2 2p 3d$. $1s^2$ and $2s^2$ are closed, so contribute no angular momentum.
For the 2p electron $l_1 = 1$ and $s_1 = \frac{1}{2}$;
for the 3d electron $l_2 = 2$ and $s_2 = \frac{1}{2}$.
$\underline{L} = \underline{l}_1 + \underline{l}_2 \Rightarrow L = 1, 2, 3$;
$\underline{S} = \underline{s}_1 + \underline{s}_2 \Rightarrow S = 0, 1$.

Combining these using all possible combinations of L and S, and the rules of vector addition, gives:

$$
\begin{array}{cccc}
 & L & S & J & \text{Level} \\
\underline{J} = \underline{L} + \underline{S} \Rightarrow & 1 & 0 & 1 & {}^1P^o_1 \\
 & 1 & 1 & 0,1,2 & {}^3P^o_0, {}^3P^o_1, {}^3P^o_2 \\
 & 2 & 0 & 2 & {}^1D^o_2 \\
 & 2 & 1 & 1,2,3 & {}^3D^o_1, {}^3D^o_2, {}^3D^o_3 \\
 & 3 & 0 & 3 & {}^1F^o_3 \\
 & 3 & 1 & 2,3,4 & {}^3F^o_2, {}^3F^o_3, {}^3F^o_4.
\end{array}
$$

Each state of an atom or ion is characterised by a unique combination of L, S and J, known as a 'level', the notation for which is explained in Sec. 4.8. Thus twelve levels arise from the configuration $1s^22s^22p3d$. Note that although some values of J appear several times, they all correspond to distinct states of the ion and it is important to retain them all.

4.7.2 *j–j coupling*

An alternative scheme for coupling angular momenta is to consider the total angular momentum, j_i, for the ith electron by combining l_i and s_i:

$$\underline{j}_i = \underline{l}_i + \underline{s}_i, \tag{4.16}$$

and then coupling these j's together to give the total angular momentum.

$$\underline{J} = \sum_i \underline{j}_i. \tag{4.17}$$

This scheme is known as *j–j coupling*. Again $J = 0$ for closed shells and sub-shells.

Worked Example: Again consider O III with configuration $1s^22s^22p3d$. For the 2p electron $l_1 = 1$ and $s_1 = \frac{1}{2}$, $\underline{j}_1 = \underline{l}_1 + \underline{s}_1$, giving $j_1 = \frac{1}{2}, \frac{3}{2}$; for the 3d electron $l_2 = 2$ and $s_2 = \frac{1}{2}$, $\underline{j}_2 = \underline{l}_2 + \underline{s}_2$, giving $j_2 = \frac{3}{2}, \frac{5}{2}$. Combining these gives:

$$
\begin{array}{cccc}
 & j_1 & j_2 & J \\
\underline{J} = \underline{j}_1 + \underline{j}_2 \Rightarrow & \frac{1}{2} & \frac{3}{2} & 1,2 \\
 & \frac{3}{2} & \frac{3}{2} & 0,1,2,3 \\
 & \frac{1}{2} & \frac{5}{2} & 2,3 \\
 & \frac{3}{2} & \frac{5}{2} & 1,2,3,4
\end{array}
$$

These correspond to exactly the same twelve J values, or levels, obtained with L–S coupling.

4.7.3 *Why two coupling schemes?*

Given that L–S and j–j coupling schemes give the same results for J, why is it necessary to have two different schemes? The answer is that the two methods are used under different circumstances.

In the non-relativistic formulation of atomic problem, as represented by the Schrödinger equation (4.1), all states with same values for L and S have the same energy. In practice, relativistic effects split this degeneracy and these effects are usually introduced as a perturbation of the non-relativistic treatment.

For light atoms, which generally mean atoms lighter than iron, relativistic effects are weak. Under these circumstances the values of L and S are approximately conserved quantities, and the L–S coupling scheme is the most appropriate.

For heavy atoms, those beyond iron for example, relativistic effects are much stronger. Under these circumstances, L and S are no longer conserved quantities and j–j coupling is more appropriate.

Light atoms are much more prevalent in astronomical spectra, so only L–S coupling is considered in any detail in this book. However it is important to remember that L and S are only approximately conserved quantities.

4.8 Spectroscopic Notation

Atoms with many active electrons can have a number of different energy levels arising from a single configuration. The splitting of configurations according to their L, S and J values requires a new method of denoting the different states. The standard notation is called *spectroscopic notation* and works within L–S coupling. Using this notation, states can be labelled as either 'terms' or 'levels'.

A 'term' is a state of a configuration with a specific value of L and S. It is denoted

$$^{2S+1}L^{(o)} .$$

The leading superscript gives the spin multiplicity, i.e. the degeneracy of the spin state. A state with $S = 0$ is a 'singlet' as $2S + 1 = 1$; a state with

$S = \frac{1}{2}$ is a 'doublet'; one with $S = 1$ is a 'triplet', and so forth. The value of the orbital angular momentum, L, is given using a capital letter using the standard letter designation (see Table 3.1). Thus S represents a state with $L = 0$, P a state with $L = 1$, D means $L = 2$, and so forth. Historically, this notation arises from analysis of the sodium spectrum (see Sec. 6.1).

The trailing superscripted 'o' in the term means 'odd parity'. Even parity terms are usually written without a superscript. The parity of a particular term, which is the same for all terms arising from a particular configuration, is defined in the following section.

Worked Example: The O III ion with configuration $1s^2 2s^2 2p3d$ can have $L = 1, 2$ or 3 and $S = 0$ or 1 (see Sec. 4.7.1). Taking all combinations of L and S, this gives rise to terms

$$^1P^o, \, ^3P^o, \, ^1D^o, \, ^3D^o, \, ^1F^o \text{ and } ^3F^o.$$

It should be noted that the splitting of configurations into terms with different energies arises even in the non-relativistic (Schrödinger) formulation. However, inclusion of relativistic effects splits these terms into levels according to their J value. A level is denoted

$$^{2S+1}L_J^{(o)},$$

where the only difference from the term notation is the addition of the subscript, J, which represents the total electron angular momentum.

Thus, for example, a $^3F^o$ term, such as the one arising from the $1s^2 2s^2 2p3d$ configuration of O III, splits into three levels with $J = 2, 3$ and 4. These levels are designated $^3F_2^o$, $^3F_3^o$ and $^3F_4^o$.

For completeness it is worth noting that for each level, characterised by a particular value of J, there are $2J + 1$ sub-levels. These are called *states* and are designated by the total magnetic quantum number M_J which takes values

$$M_J = -J, \, -J + 1, \ldots, J - 1, J.$$

These states are degenerate in the absence of an external field. The splitting of levels into states in a magnetic field is generally known as the *Zeeman effect*.

4.9 Parity of the Wavefunction

The parity of the wavefunction is determined by how the wavefunction behaves upon inversion. Inversion is the operation of reflecting the wavefunction through the origin, here the atomic nucleus, and is equivalent to replacing vector $\underline{r}$ with $-\underline{r}$. Given the symmetry of the atom, the square of the wavefunction, i.e. the probability distribution of the electrons, must be unchanged by this operation. Neglecting spin, this means

$$\psi(\underline{r}_1, \underline{r}_2, \ldots, \underline{r}_N) = \pm\psi(-\underline{r}_1, -\underline{r}_2, \ldots, -\underline{r}_N). \tag{4.18}$$

Even parity states are given by $+\psi$ and odd parity states are given by $-\psi$.

In practice the parity of all terms and levels arising from a particular configuration can be determined simply by summing the orbital angular momentum quantum numbers for each of the electrons. With this simple rule, the parity is given by

$$(-1)^{l_1+l_2+\cdots l_N}. \tag{4.19}$$

As closed shells and sub-shells have an even number of electrons, it is again only necessary to explicitly consider the active electrons.

Thus for the O III configuration $\mathrm{ls}^2 2\mathrm{s}^2 2\mathrm{p} 3\mathrm{d}$, it is only necessary to consider the sum $l(2\mathrm{p}) = 1$ and $l(3\mathrm{d}) = 2$. This gives $(-1)^{1+2} = -1$, which explains why all the terms and levels arising from this configuration were all labelled odd above.

The parity of a configuration is important since it leads to a rigorous dipole selection rule known as the Laporte rule. The Laporte rule states:

All electric dipole transitions connect states of opposite parity.

In other words (strong) transitions can only link configurations with even to those with odd parity, and *vice versa*.

4.10 Terms and Levels in Complex Atoms

A single configuration can lead to several terms. These terms have different energies. It is worth considering a few examples.

Example 1: The helium atom.

(1) The ground state is $1\mathrm{s}^2$.
   This is a closed shell, with $L = 0$ and $S = 0$, hence it gives rise to a single, even parity term $^1\mathrm{S}$, and level $^1\mathrm{S}_0$.

(2) The first excited configuration is 1s2s.

This has $l_1 = l_2 = 0$ and hence $L = 0$,

but $s_1 = s_2 = \frac{1}{2}$ giving both $S = 0$ (singlet) or $S = 1$ (triplet) states. The energy ordering of atomics states is given by Hund's rules. Hund's first rule governs ordering of terms with different spin multiplicities:

For a given configuration, the state with the maximum spin multiplicity is lowest in energy.

So the ^3S term ($^3S_1$ level) is lower in energy than the ^1S term ($^1S_0$ level). In practice the splitting between these terms is 0.80 eV.

(3) The next excited configuration is 1s2p, which has odd parity.

This has $l_1 = 0$ and $l_2 = 1$, giving $L = 1$;

again $s_1 = s_2 = \frac{1}{2}$, giving both $S = 0$ and $S = 1$ terms.

Following the rule above, the $^3P^o$ term is lower than the $^1P^o$ term, in this case by 0.25 eV. The $^3P^o$ is also split into three levels: $^3P_0^o$, $^3P_1^o$ and $^3P_2^o$.

Figure 5.2 depicts the energy levels of helium in a form known as a Grotrian diagram. The layout of these diagrams is discussed in Sec. 5.4.

Example 2: The carbon atom.

Start by considering the excited state configuration $1s^2 2s^2 2p3p$.

It is only necessary to consider the outer two electrons for which:

$$l_1 = 1, \quad s_1 = \frac{1}{2},$$
$$l_2 = 1, \quad s_2 = \frac{1}{2}.$$

These give $L = 0, 1, 2$ and $S = 0, 1$, which give rise to the following terms, all with even parity: ^1S, ^3S, ^1P, ^3P, ^1D and ^3D.

Now consider the ground state configuration of carbon $1s^2 2s^2 2p^2$.

This configuration also has $l_1 = 1, s_1 = \frac{1}{2}$ and $l_2 = 1, s_2 = \frac{1}{2}$.

However the Pauli Principle restricts which terms are allowed. For example the term ^3D includes the state:

$$\left(l_1 = 1, m_1 = +1, s_1 = \frac{1}{2}, s_{1z} = +\frac{1}{2} \right)$$
$$\left(l_2 = 1, m_2 = +1, s_2 = \frac{1}{2}, s_{2z} = +\frac{1}{2} \right)$$

which is allowed when $n_1 = 2, n_2 = 3$, but is forbidden when $n_1 = 2, n_2 = 2$ by the Pauli exclusion principle since both electrons have precisely the same quantum numbers.

There are general methods of determining which terms are allowed for a configuration with a multiply occupied open shell [see Bransden and Joachain (2003) in further reading]. However there is a rule of thumb which suffices for present purposes. It turns out that for systems with equivalent electrons, that is, electrons which have the same n and l values, then the sum $L + S$ for these electrons must be even for the Pauli Principle to be satisfied.

The ground state configuration of carbon, C I, thus gives terms ^1S, ^3P and ^1D. The ^3P term has the highest spin and is thus the ground state term. The other two terms have however, the same spin multiplicity, so which is lower in energy? Hund's second rule states:

For a given configuration and spin multiplicity, the state with the maximum orbital angular momentum is the lowest in energy.

In the case of the ground state configuration of carbon, the ^1D state lies 1.42 eV lower in energy than the ^1S state, but is 1.26 eV above the ^3P state.

The examples above have only considered terms and have neglected splitting according to J where it arises. This fine-structure splitting becomes increasingly important for high Z (i.e. heavy) atoms.

The ^3P ground state of carbon has $L = 1$ and $S = 1$, which lead to $J = 0, 1, 2$. Allowed levels are thus $^3P_0$, $^3P_1$ and $^3P_2$.

The energy order of these is given by Hund's third rule:

The lowest energy is obtained for lowest value of J in the normal case and for highest J value in the inverted case.

The normal case is a shell which is less than half filled, for example $2p^2$ as in carbon. The inverted case is a shell which is more than half full such as the $2p^4$ found in the ground state of atomic oxygen, which also has a ^3P ground state.

Thus for carbon one gets the energy order:

$$^3P_0 < {}^3P_1 < {}^3P_2 ,$$

whereas for oxygen, one gets

$$^3P_2 < {}^3P_1 < {}^3P_0 .$$

It should be noted that this situation does not arise for configurations with half-filled shells since the lowest energy term, the only one for which

Hund's rules apply rigorously, always has $L = 0$ so that there is only a single level.

To summarise the energy ordering of levels given by Hund's rules:

(1) For a given configuration, the term with maximum spin multiplicity lies lowest in energy;
(2) For a given configuration and spin multiplicity, the term with the largest value of L lies lowest in energy;
(3) For atoms with less than half-filled shells, the level with the lowest value of J lies lowest in energy;
(4) For atoms with more than half-filled shells, the level with the highest value of J lies lowest in energy.

It should be noted that Hund's rules are only applicable within $L–S$ coupling. Furthermore, they are only rigorous for ground states. However they are in practice almost always followed for all states of atoms and ions. They are therefore also useful for determining the energy ordering of excited states.

Problems

4.1 What is the ground state configuration, term and level of the beryllium atom, Be? One of the outer electrons in Be is promoted to the 3d orbital. What terms and levels can this configuration have?

4.2 An excited helium atom has the configuration $3d^2$. What values of the total orbital angular momentum quantum number, L, and total spin angular momentum quantum number, S, are allowed? Use spectroscopic notation to give the terms which arise from the combinations of L and S allowed under the Pauli exclusion principle. Use Hund's rules to suggest the energy order in which these terms are likely to occur. For each term deduce the allowed values of the total angular momentum quantum number, J. Hence give the full designation of each level, including parity, using spectroscopic notation. Which level has the lowest energy?

4.3 The symbol for a particular level is quoted as $^4F^o_{\frac{7}{2}}$. What are the values of L, S and J for this level? How many states does it have? What are the other levels for this term?

Use the value of S to determine the minimum number of electrons that could give rise to this term. Suggest a possible configuration that could give this term.

4.4 Symbols for particular levels of three different atoms are written as $^1D_1$, $^0D_{\frac{5}{2}}$ and $^3P_{\frac{3}{2}}$. Explain in each case why the symbol must be wrong.

4.5 The lithium atom Li has three electrons. For each of the following configurations, what terms will be present: (a) $1s^22p$, (b) $1s2s3s$, and (c) $1s2p3p$?

4.6 The $1s^22s^22p^63s^23p^63d5p$ configuration of Ca is formed during recombination. Derive the terms that arise from this configuration and explain which you would expect to be lowest in energy. What levels can arise from the lowest energy term?

CHAPTER FIVE

HELIUM SPECTRA

'Born of the Sun.'

– Stephen Spender, *I Think Continually* (1933)

5.1 He I and He II Spectra

Helium is the second-most abundant element after hydrogen comprising more than 25% of the Universe's atomic matter by weight. Helium can exist in its atomic form, He I, as singly ionised He II, or as doubly ionised He III. He III is of course the base helium nucleus He^{2+}, which is also the α particle formed during radioactive decay, and which has no spectrum.

The He^+ ion is a one-electron system. It therefore has a hydrogen-like spectrum except that the binding of the energy levels and the transition frequencies are scaled by a factor of Z^2, where for helium, $Z = 2$.

Thus for example, the $n = 2-1$ Lyα for hydrogen is observed at a wavelength of 1216 Å. The corresponding transition in He II, which is also referred to as Lyα, is observed at 304 Å. This transition is observable in absorption in the interstellar medium (ISM) where it is difficult to measure helium abundances by other means. Such observations have been made from the extreme ultraviolet explorer (EUVE) satellite.

As discussed in Chapter 3, hydrogen lines occur at

$$\frac{1}{\lambda} = R_{\mathrm{H}} \left(\frac{1}{n_1^2} - \frac{1}{n_2^2} \right) . \qquad (5.1)$$

For the Balmer series, $n_1 = 2$, $n_2 = 3, 4, 5, \ldots$. For He II, similar lines in the visible occur at

$$\frac{1}{\lambda} = 4R_{He}\left(\frac{1}{n_1^2} - \frac{1}{n_2^2}\right) = R_{He}\left[\frac{1}{\left(\frac{n_1}{2}\right)^2} - \frac{1}{\left(\frac{n_2}{2}\right)^2}\right], \qquad (5.2)$$

where R_{He} is the Rydberg constant for He^+, which is intermediate in value between R_H and R_∞ (see Sec. 3.7). Lines with $n_1 = 4$ and $n_2 = 6, 8, 10, \ldots$ are therefore almost coincident with the Balmer series. The shift due to the reduced mass factor in the Rydberg constant can only be seen at high resolution. However, intermediate lines with $n_2 = 5, 7, 9, \ldots$ are also observed. It was originally thought that this spectrum, which is known as the Pickering series, was due to atomic hydrogen with half-integer quantum numbers! The Pickering series can be observed in the spectra of O-stars. Indeed the classification of O-stars depends on the relative strength of the absorption by He II and He I in their atmosphere (see Fig. 5.1).

Emission lines of He II are also observed among the recombination line series in a nebulae. However these emissions only come from the

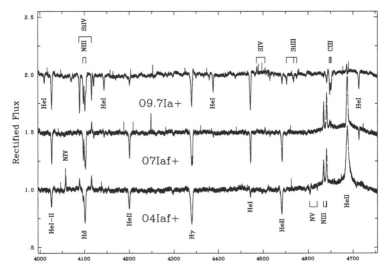

Fig. 5.1. Spectra of early (O4 star HDE 269698), mid (O7 star AzV 232) and late (O9 star Sk-66°169) type O-supergiant stars showing the sensitivity of the subtypes to the ratio of He I 4471 Å to He II 4542 Å absorptions. The spectra were recorded using the Very Large Telescope (VLT) by P.A. Crowther *et al.* [see *Astrophys. J.* **579**, 774 (2002)]. (P.A. Crowther, private communication.)

region of nebula close to the central star where there are sufficient high-energy, ultraviolet photons to ionise He^+. He^+ requires 54.4 eV, compared to 13.6 eV which is needed to ionise H. He I recombination spectra also come from a smaller physical region than the H II region as the ionisation potential of He is 24.6 eV.

Strong He I emissions are observed from nebulae including the lines:

$$4471\,\text{Å} \quad 1s2p - 1s4d\ ^3P^o - {}^3D,$$
$$5875\,\text{Å} \quad 1s2p - 1s3d\ ^3P^o - {}^3D,$$
$$6678\,\text{Å} \quad 1s2p - 1s3d\ ^1P^o - {}^1D.$$

The 5875 Å line lies in the yellow and absorbs in the Sun. Observation of this previously unknown line in the Sun during the solar eclipse of 1868 led Sir James Lockyer (1836–1920) to postulate the presence of a new element: helium. He was widely ridiculed for his boldness but was vindicated when Sir William Ramsay (1852–1916) isolated helium in his laboratory at University College London more than twenty years later.

Like hydrogen, transitions involving the ground state of He lie far in the ultraviolet. For example, the He I $1s^2 - 1s2p$ $^1S - {}^1P^o$ transition lies at 584 Å. This transition is known as the *He I resonance line*. The phrase 'resonance line' is used to denote the longest wavelength, dipole-allowed transition arising from the ground state of a particular atom or ion.

5.2 Selection Rules for Complex Atoms

Strong transitions are driven by electric dipoles. Electric dipole selection rules are of two types: rigorous rules which must always be obeyed, and propensity rules which, when violated, lead to weaker (forbidden) transitions.

The rigorous selection rules which all electric dipole transitions, even so-called forbidden ones, must satisfy are:

(1) ΔJ must be 0 or ± 1 with $J = 0 \leftrightarrow 0$ forbidden.
(2) $\Delta M_J = 0, \pm 1$.
(3) Parity changes i.e. even $\leftrightarrow$ odd. This is the Laporte rule (Sec. 4.9).

All electric dipole transitions must obey these rules. There is an additional, and more complex set of rules, which is satisfied by systems with a single electron but not rigorously by complex atoms. Failure to satisfy these rules, which I will call propensity rules, does not completely rule out a transition but generally makes it much weaker.

The propensity rules for electric dipole transitions, which lead to stronger transitions if satisfied, are:

(4) The spin multiplicity is unchanged, $\Delta S = 0$.

(5) Only one electron jumps, i.e. the configuration of the two states must differ by only the movement of a single electron. This movement is governed by the rules Δn any and $\Delta l = \pm 1$, thus:

$$2s^2 \leftrightarrow 2s2p \text{ is allowed};$$
$$2s^2 \leftrightarrow 2s3d \text{ or } 2s^2 \leftrightarrow 3p^2 \text{ is forbidden}.$$

Configuration interaction (CI) weakens this rule since CI means that the real wavefunction is a (small) mixture of different configurations. For example, the ground state of Be, whose configuration is usually written $1s^2 2s^2$, is more precisely represented by 95% $1s^2 2s^2$ mixed with about 5% of the same symmetry configuration $1s^2 2p^2$.

(6) $\Delta L = 0, \pm 1, L = 0 \leftrightarrow 0$ forbidden, thus:

$$^1S \leftrightarrow {}^1P^o \text{ or } {}^3D \leftrightarrow {}^3P^o \text{ are allowed};$$
$$^1S \leftrightarrow {}^1S^o \text{ or } {}^3S \leftrightarrow {}^3D^o \text{ are forbidden}.$$

Note that it is possible to change l and not L, for example, 2p3p $^1P \leftrightarrow$ 2p3d $^1P^o$ is allowed. Such a transition is observed for C I.

A full list of selection rules are given in Table 5.1.

Electric dipole transitions which satisfy all the rigorous selection rules as well as the propensity rules are referred to as **allowed** transitions. These transitions are generally strong and have Einstein A coefficients which are typically bigger than $10^6 \, \text{s}^{-1}$.

Photons do not change spin, so transitions usually occur between terms with the same spin state, as expressed by the rule $\Delta S = 0$. However, relativistic effects mix spin states, particularly for high Z atoms or ions. As a result of relativistic effects one can get (weak) spin changing transitions; these are called **intercombination lines**. Intercombination lines are denoted by one square bracket, for example:

$$\text{C III}] \, 2s^2 \, {}^1S - 2s2p \, {}^3P^o \text{ at } 1908.7 \, \text{Å}.$$

This transition is important because the C^{2+} 2s2p $^3P^o$ state is metastable, i.e. it has no allowed radiative decay so that this transition determines the

Table 5.1. Selection rules for atomic spectra. Rules 1, 2 and 3 must always be obeyed. For electric dipole transitions, intercombination lines violate rule 4 and forbidden lines violate rule 5 and/or 6. Electric quadrupole and magnetic dipole transitions are also described as forbidden.

	Electric dipole	Electric quadrupole	Magnetic dipole
1.	$\Delta J = 0, \pm 1$ Not $J = 0 - 0$	$\Delta J = 0, \pm 1, \pm 2$ Not $J = 0 - 0, \frac{1}{2} - \frac{1}{2}, 0 - 1$	$\Delta J = 0, \pm 1$ Not $J = 0 - 0$
2.	$\Delta M_J = 0, \pm 1$	$\Delta M_J = 0, \pm 1, \pm 2$	$\Delta M_J = 0, \pm 1$
3.	Parity changes	Parity unchanged	Parity unchanged
4.	$\Delta S = 0$	$\Delta S = 0$	$\Delta S = 0$
5.	One electron jumps Δn any $\Delta l = \pm 1$	One or no electron jumps Δn any $\Delta l = 0, \pm 2$	No electron jumps $\Delta n = 0$ $\Delta l = 0$
6.	$\Delta L = 0, \pm 1$ Not $L = 0 - 0$	$\Delta L = 0, \pm 1, \pm 2$ Not $L = 0 - 0, 0 - 1$	$\Delta L = 0$

lifetime of this state. Actually, the situation is more subtle than this. The $^3P^o$ term splits into three levels: $^3P^o_0$, $^3P^o_1$ and $^3P^o_2$. The electric dipole intercombination line at 1908.7 Å is actually $^1S_0 - {}^3P^o_1$. It has an A value of 114 s^{-1}.

The transition $^1S_0 - {}^3P^o_2$, which occurs at 1906.7 Å, is completely forbidden by dipole selection rules as $\Delta J = 2$. It only occurs via a very weak magnet quadrupole transition. The 1906.7 Å line is 10^5 times weaker than the already-weak line at 1908.7 Å; it has an A value of 0.0052 s^{-1}. These two lines can be used to give information on the electron density, as discussed in Sec. 7.1. Finally the transition $^1S_0 - {}^3P^o_0$ is a $J = 0 - 0$ transition, which is completely forbidden by both dipole and quadrupole selection rules. This transition is not observed.

Electric dipole transitions which violate the propensity rules 5 and/or 6 are called **forbidden transitions**. These are labelled by square brackets. For example,

$$1906.7 \text{ Å } [\text{C \textsc{iii}}] \ 2s^2 \ {}^1S_0 - 2s2p \ {}^3P^o_2 ;$$

$$322.57 \text{ Å } [\text{C \textsc{iii}}] \ 2s^2 \ {}^1S_0 - 2p3s \ {}^1P^o_2 ;$$

are both forbidden lines of C^{2+}. The former is a magnetic transition while the latter is an electric dipole transition involving the movement of two electrons. Forbidden transitions are generally weaker than intercombination lines.

It is also possible to get transitions driven by higher electric multi-poles or magnetic moments. The only important ones of these are electric quadrupole and magnetic dipole transitions. The selection rules for these transitions are also given in Table 5.1. Even when all the rules are satisfied, electric quadrupole and magnetic dipole transitions are both much weaker than the allowed electric dipole transitions. They are thus also referred to as forbidden transitions.

Typical lifetimes, that is inverse Einstein A coefficients, for allowed decays via each mechanism are

$$\tau_{\text{dipole}} \sim 10^{-8}\text{s}, \quad \tau_{\text{magnetic}} \sim 10^{-3}\text{s}, \quad \tau_{\text{quadrupole}} \sim 1\text{s}.$$

These timescales mean that states only decay by forbidden transitions when there are no decay routes via allowed transitions.

Finally it should be noted that even the rigorous selection rules given above can be modified when nuclear spin effects are taken into consideration. These result in rigorous selection rules for electric dipole transitions based on the final angular momentum. In particular:

$$\Delta F \text{ must be } 0 \text{ or } \pm 1 \text{ with } F = 0 \leftrightarrow 0 \text{ forbidden}.$$

It is only very rarely necessary to consider this.

5.3 Observing Forbidden Lines

States decaying only via forbidden lines live for a long time on an atomic, if not an astronomical, timescale. Such states are called *metastable states*.

Forbidden lines are often difficult to study in the laboratory as collision-free conditions are needed to observe metastable states. In this context it must be remembered that laboratory ultrahigh vacuums are significantly denser than so-called dense interstellar molecular clouds. Astrophysically, low density environments are common. In these environments the time between collisions is very long and an atom in an excited state has time to radiate even when it is metastable.

Emissions due to forbidden lines are important in hot, low density regions such as H II regions, planetary nebulae, the solar corona and the Earth's aurora. Observing them gives direct information on the populations of excited levels. As the transitions are weak they have low optical depth and therefore give reliable population information.

Forbidden transitions are usually only important when an excited state cannot decay via an allowed transition, i.e. when this state is metastable. For this reason, forbidden lines are normally only important for low-lying states since higher states nearly always have possible radiative decay routes via allowed transitions. For neutral atoms or ones with low ionisation, this means that forbidden transitions are often observed in the infrared. Of course, this wavelength shifts with the level of ionisation so that forbidden lines for ions occur throughout the visible, ultraviolet and even at X-ray wavelengths (see Sec. 5.6).

One common source of forbidden transitions in the infrared arises from the relaxation of excited terms within the ground state configuration and levels within the ground state term. Such transitions are completely dipole-forbidden by the Laporte rule and hence undergo weak magnetic transitions instead.

5.4   Grotrian Diagrams

The German astrophysicist Walter Grotrian (1890–1954) invented a simple diagrammatic means of representing the many states and transitions of atoms and ions. These are called Grotrian diagrams and several of them can be found in this book.

The standard structure of a Grotrian diagram is as follows:

(1) The vertical scale is energy. It starts from the ground state at zero, and extends to the first ionisation limit. Sometime the binding energy, expressed relative to the first ionisation limit, is given as the right-hand vertical scale. Terms (levels) are represented by horizontal lines.

(2) States with the same term, or sometimes the same level if fine structure effects are large, are stacked vertically and labelled by the principal quantum number n of the outer electron. For example, He I singlet states (see Fig. 5.2), with configuration $1snd$ appear in the column headed ^1D and are labelled by $n = 3, 4, 5, \ldots$.

(3) Terms are grouped by spin multiplicity.

(4) States are linked by observed transitions with numbers giving the wavelength of the transition, usually as an integer, in Å. Thicker lines denote stronger transitions and forbidden transitions are given by dashed lines.

There are some variations on this structure of the diagram according to the composer and the system being considered. Grotrian diagrams

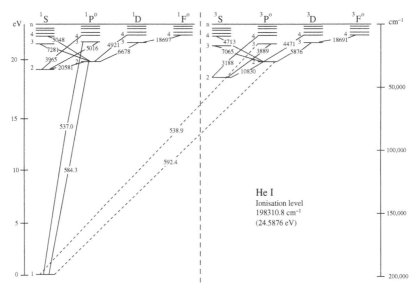

Fig. 5.2. Grotrian diagram for He I. The running numbers denote the principal quantum number of the active electron. The left-hand side of the figure is for 'para' singlet helium and the right-hand side is for 'ortho' triplet helium.

are useful for getting a feel for a particular spectrum but not suitable for detailed or high accuracy work. A particularly comprehensive and useful series of Grotrian diagrams for each element and ionisation stage has been compiled by Bashkin and Stoner (see further reading).

Figure 5.2 gives the Grotrian diagram for He I. This diagram splits into two: the left-hand side gives singlet ('Para') He, while the right corresponds to triplet ('Ortho') He. The terms ortho and para, short for 'orthodox' and 'paradox', arise from the spin multiplicity of the atom with the ortho or orthodox states being three, $2S + 1 = 3$, times as likely as the $2S + 1 = 1$ para or paradox states. This terminology is not used for other atomic systems but has been adopted for molecular spectra where it is the nuclear rather than the electron spin multiplicities which give rise to differing weights in the spectrum.

The ortho and para states of He I are linked by very weak intercombination lines, two of which are shown in Fig. 5.2. In particular, the 2 ^{3}S state can only decay to the 1 ^{1}S ground state. This transition is dipole-forbidden by the Laporte rule. It decays via a magnetic dipole transition. The transition is weaker than an allowed magnetic dipole transition since it is both spin-forbidden, as $\Delta S = 1$, and involves an electron jump. It has $A = 1.2 \times 10^{-4}\,\text{s}^{-1}$ which means the helium 1s2s ^{3}S state has a lifetime of 8000 s.

5.5 Potential Felt by Electrons in Complex Atoms

In the orbital approximation, each electron moves in an effective potential, $V_i(r_i)$ (see Sec. 4.2). V_i is complicated but aspects of its general form can be understood by considering its behaviour as $r_i \to 0$ and as $r_i \to \infty$.

Consider an atom or ion with N electrons and nuclear charge $+Ze$. For simplicity assume that $N \leq Z$. As the coordinate of the ith electron gets very large ($r_i \to \infty$), there are $N - 1$ electrons left near the nucleus. In this regime the potential felt by this outer electron is given by:

$$V_i(r_i) \overset{r \to \infty}{\longrightarrow} -\frac{Z}{r_i} + \frac{(N-1)}{r_i} = -\frac{(Z-N+1)}{r_i}. \tag{5.3}$$

For a neutral atom $N = Z$, and $V_i(r_i) \to -r_i^{-1}$. Conversely as the ith electron gets very close to the nucleus ($r_i \to 0$), the electron moves inside all the other electrons and therefore feels the full nuclear charge.

$$V_i(r_i) \overset{r_i \to 0}{\longrightarrow} -\frac{Z}{r_i}. \tag{5.4}$$

This analysis does not solve the problem of the form of $V_i(r_i)$ but it does give its limits:

$$-\frac{Z}{r} < V(r) < -\left(\frac{Z-N+1}{r}\right). \tag{5.5}$$

This is shown schematically in Fig. 5.3. It is tempting to interpolate between the two limits shown on the figure but this would be wrong. In practice, the shell structure of the atom introduces fluctuations into the potential felt by the individual electrons at intermediate values of r.

If the outermost electron of an atom or ion lay entirely outside the other electrons it would be quasi-hydrogenic and its binding energy would be given by the expression

$$E_{nl} \simeq -R_\infty \frac{Z_{\text{eff}}^2}{n^2}, \quad Z_{\text{eff}} = Z - N + 1. \tag{5.6}$$

However penetration makes the potential more attractive. As discussed in Sec. 4.4 and illustrated by Fig. 3.1, penetration is the ability of an outer electron to penetrate the charge cloud of the inner electrons. This can be

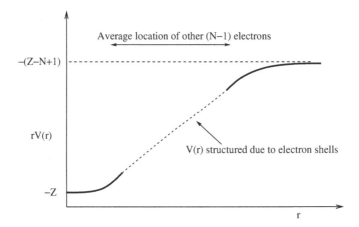

Fig. 5.3. Approximate shape of the potential felt by a single electron in a complex atom with N electrons. The vertical axis represents $rV(r)$.

thought of as raising the value of Z_{eff} felt by the outer electron, but it is better to think of penetration as lowering the principal quantum number n:

$$E_{nl} \simeq -R_\infty \frac{Z_{eff}^2}{\nu_{nl}^2},\tag{5.7}$$

where ν is known as the effective quantum number. It is useful to define the effective quantum number in terms of n, and something called the quantum defect, μ_{nl}, via the expression

$$\nu_{nl} = n - \mu_{nl}.\tag{5.8}$$

As s orbitals penetrate more than p, p orbitals penetrate more than d, and so forth, one finds in general that

$$E_{no} < E_{n1} < E_{n2}\dots,$$

which means that

$$\mu_{no} > \mu_{n1} > \mu_{n2}\dots.$$

Quantum defects are discussed in more detail in Sec. 6.1.

5.6   Emissions of Helium-Like Ions

Any ion with only two electrons is known as a helium-like ion. It turns out that spectra of helium-like or K shell ions have become an important diagnostic because their characteristic spectra can give important physical information about a hot environment. If a helium-like ion has one electron excited to its $n = 2$ level, then the various states, which are all fairly close in energy, can decay by the following routes:

W $1s2p - 1s^2\ {}^1P_1^o - {}^1S_0$, resonance line,

X $1s2p - 1s^2\ {}^3P_2^o - {}^1S_0$, magnetic quadrupole line,

Y $1s2p - 1s^2\ {}^3P_1^o - {}^1S_0$, intercombination line,

Z $1s2s - 1s^2\ {}^3S_1 - {}^1S_0$, forbidden transition.

The letters are the standard labels for these transitions, an example of which is given for O VII in Fig. 5.4; note that lines X and Y are close together and are usually not separately resolved.

Careful monitoring of the relative intensities of these lines can give information both on the temperature and density of the environment and the ionisation mechanism involved. In particular the critical density n_c (see Sec. 2.6) of the transitions increases strongly with the nuclear charge, Z. So, for example, n_c of Si XIII is about 10000 times greater than n_c of the isoelectronic C V. This means that a large range of electron densities can be studied by monitoring a variety of helium-like ions.

Figure 5.5 gives a simple Grotrian diagram for O VII which is helium-like, and one of the species commonly observed. Its spectrum is also observed in solar flares. The O VII resonance line lies at 21.6 Å, which is in the X-ray region of the spectrum. At these wavelengths the transitions depicted in the figure appear as a triplet as the intercombination doublet is usually not resolved. These transitions lie close enough together for their spectra to be recorded simultaneously by instruments on board satellites such as the XMM-Newton. X-ray spectra are discussed in Chapter 8 where Fig. 8.3 shows emission from He-like iron or Fe XXV.

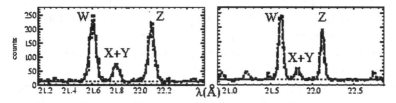

Fig. 5.4. Helium-like oxygen, O VII, triplet X-ray spectra from the coronae of the stars Procyon (*left*) and Capella (*right*). See the text for an explanation of the labels for each line. Spectra were recorded using the satellite Chandra. [Adapted from S.M. Kahn *et al.*, Philos. Trans. R. Soc. Lond., Ser **A 360**, 1923 (2002).]

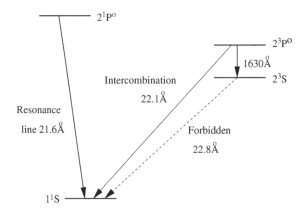

Fig. 5.5.   Term diagram showing for helium-like oxygen, O VII, showing transitions from the 1s2*l* states.

Problems

5.1 A transition of hydrogen-like helium, ^{4}He$^+$, is observed close to hydrogen Hα. Between which states of He$^+$ is this transition? Estimate the wavenumber of this transition, stating any assumptions made.

5.2 Emission of He II are observed at 1640 Å. To what transitions does this radiation correspond to? What other emission lines must also be present in the spectrum?

5.3 A nebula is composed only of hydrogen and helium, and consists of four shells around a hot, central star. In shell A, closest to the star, all atoms are fully ionised; in shell B, helium and hydrogen are both singly ionised; shell C is an H II region with neutral helium; shell D, which is furthest from the central star, is an H I region. Explain what (if any) recombination line spectra you would expect to observe from each region.

5.4 The helium-like ion N VI is formed in the hot intergalactic medium with configurations 1s2s and 1s2p. What levels can be formed from these configurations? By what mechanism would you expect each level to emit to the 1s^2 ground state? Order each of these transitions according to its approximate strength.

5.5 The lithium atom, Li, has three electrons. Consider the following configurations of Li: (a) 1s^{2}2p, (b) 1s2s3s, (c) 1s2p3p. By considering the configurations only, state which of the three sets of transitions between the configurations (a), (b) and (c) are electric dipole allowed and electric dipole forbidden transitions?

CHAPTER SIX

ALKALI ATOMS

Lithium, sodium, potassium and rubidium all have ground state electronic structures which consist of one electron in an s orbital outside a closed shell. This single 'optically active' electron gives these atoms, the alkali metals, similar chemical behaviour and fairly simple spectra. Even so, the presence of the inner or core electrons lead to a number of complications which are not present in the spectrum of simple one-electron atoms.

6.1 Sodium

Sodium, Na, has $Z = 11$ and a ground state configuration of $1s^2 2s^2 2p^6 3s^1$. If the outer 3s electron was completely screened then it would feel an effective nuclear charge $Z_{eff} = 11 - 10 = 1$ and its energy levels would obey the Rydberg formula. In practice, the 3s electron penetrates and reduces the effective quantum number of the electron giving a revised formula:

$$E_{nl} = -R_\infty \frac{Z_{eff}^2}{(n - \mu_{nl})^2},$$
(6.1)

where μ_{nl} is the quantum defect. The quantum defect formula was originally proposed by Rydberg. Unlike his hydrogenic formula, Eq. (3.9), this depends on l as the degree of electron penetration is l-dependent: in particular, electrons with low l penetrate more and hence have lower energy. An advantage of this form is that usually μ_{nl} only depends weakly on n, allowing values of the quantum defect to be transfered between states with different n but the same l. Table 6.1 gives values of the quantum defect for low-lying states of sodium. Note the $\mu_{43} \approx 0$ means that there is effectively no penetration for f electrons; this is also true for electrons with l greater

81

Table 6.1.   Quantum defects μ_{nl} for sodium.

l		$n = 3$	$n = 4$	$n = 5$	$n = 6$	$n = \infty$
0	s	1.373	1.357	1.353	1.351	1.348
1	p	0.883	0.867	0.862	0.857	0.855
2	d	0.012	0.013	0.014	0.014	0.015
3	f	—	0.000	0.000	0.000	0.000

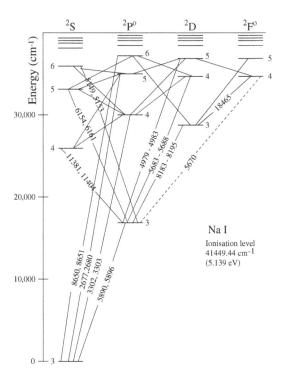

Fig. 6.1.   Grotrian diagram for sodium. Transitions are labelled with more than one wavelength due to the effects of fine structure.

than 3. Use of the quantum defects for sodium gives a hydrogen-like level structure with splitting on l (and J) see Fig. 6.1.

Worked Example: The spectrum of S VI, sodium-like sulphur, shows a series of ns–3p transitions with the series limit at $604310\,\mathrm{cm}^{-1}$. The first two transitions in this series lie at $257109\,\mathrm{cm}^{-1}$ and $398238\,\mathrm{cm}^{-1}$. Estimate where the third transition in the series should lie.

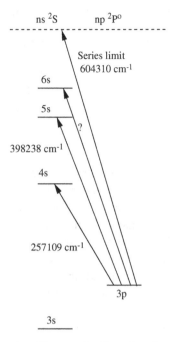

Fig. 6.2. Simplified Grotrian diagram for S VI showing transitions discussed in the worked example.

This data is best shown diagrammatically as can be seen from Fig. 6.2. Using this figure, the energy of the 4s level, $E(4s)$ is $257109 - 604310 = -347201 \, \text{cm}^{-1}$ and $E(5s) = 398238 - 604310 = -206072 \, \text{cm}^{-1}$. For the outer electron in S^{5+}, $Z_{\text{eff}} = 6$, so the quantum defects are given by the expression

$$\mu(ns) = n - 6 \sqrt{\frac{R_\infty}{-E(ns)}}.$$

Using the values of $E(4s)$ and $E(5s)$ determined above gives $\mu(4s) = 0.627$ and $\mu(5s) = 0.622$. Assuming $\mu(6s) \simeq 0.62$ gives $E(6s) = -136487 \, \text{cm}^{-1}$ and hence the 6s–3p transition at $467823 \, \text{cm}^{-1}$. In practice, this transition is observed at $467949 \, \text{cm}^{-1}$. Note that $E(ns)$ is negative since it is binding energy which goes to zero at the ionisation limit.

The Grotrian diagram for sodium, Fig. 6.1, shows that transitions with $\Delta l = \pm 1$ and $\Delta L = \pm 1$ dominate the spectrum. However, as implied by the multiple wavelengths given for most transitions, these lines are actually split into components.

Table 6.2.   Spectral series of sodium.

Series name	Transitions	n values	Multiplicity
Sharp	$n\,^2S_{\frac{1}{2}} \rightarrow 3\,^2P^o_{\frac{3}{2},\frac{1}{2}}$	$n = 4,5,6,\ldots$	doublets
Principal	$n\,^2P^o_{\frac{3}{2},\frac{1}{2}} \rightarrow 3\,^2S_{\frac{1}{2}}$	$n = 3,4,5,\ldots$	doublets
Diffuse	$n\,^2D_{\frac{5}{2},\frac{3}{2}} \rightarrow 3\,^2P^o_{\frac{3}{2},\frac{1}{2}}$	$n = 3,4,5,\ldots$	triplets
Fundamental	$n\,^2F^o_{\frac{7}{2},\frac{5}{2}} \rightarrow 3\,^2D_{\frac{5}{2},\frac{3}{2}}$	$n = 4,5,6,\ldots$	triplets

Consider the spectrum of sodium and, in particular, emissions into the $n = 3$ shell of sodium. The spectral series, most of whose transitions lie in the visible, are given in Table 6.2. These series are named after their spectral appearance rather than their discoverers. Note the initial letters of the names of each series; these gave rise to the use of the notations s, p, d and f for orbital angular momentum states $l = 0, 1, 2$ and 3 respectively, which of course, are the corresponding orbital angular momenta of the emitting states. Orbitals with $l > 3$ are simply labelled alphabetically, with the omission of the letter j.

As noted in Table 6.2, each member of each series is split into more than one line. These splittings are large enough to be resolved with a high-resolution spectrograph and, as discussed later in this chapter, are very important observationally. However before discussing their astronomical importance it is necessary to understand the physics that gives rise to the different multiplets, and in particular why some series give two lines (doublets) and others three lines (triplets).

6.2   Spin-Orbit Interactions

Both the orbital angular momentum L and spin angular momentum S give rise to an internal magnetic field within the atom. It is the magnetic interactions between the spin and orbital motions of the electrons which split terms into levels.

The most important interaction is between the spin magnetic moment of an electron which is given by

$$\underline{\mu}_s = -2\frac{\mu_B}{\hbar}\hat{s}, \tag{6.2}$$

and the magnetic field due to the orbital motion of the electron. This is the only source of spin-orbit interaction considered here.

Equation (6.2) defines the spin magnetic moment in terms of the Bohr magneton μ_B, where

$$\mu_B = \frac{e\hbar}{2m_e} = \frac{1}{2} \text{ a.u.} = 9.27 \times 10^{-24} \text{ JT}^{-1}.$$

The Bohr magneton is the basic unit, or quantum, of magnetic moment.

The magnetic field due to the orbital motion of an electron is

$$\underline{B} = \frac{\hbar^2}{2} \frac{\underline{v} \times \underline{r}}{c^2 r} \frac{dV}{dr}, \tag{6.3}$$

where $V(r)$ is the potential experienced by the electron. As the orbital angular momentum can be defined in terms of the velocity $\underline{v}$ by the expression

$$\underline{l} = m_e \underline{r} \times \underline{V}, \tag{6.4}$$

$$\underline{B} = -\frac{\hbar^2}{2} \frac{\underline{l}}{m_e r c^2} \frac{dV}{dr}. \tag{6.5}$$

The factor of c^{-2} means that the resulting magnetic field, and hence the energy shift, is small.

The energy shift due to interactions is given by $-\underline{\mu} \cdot \underline{B}$. The spin-orbit interaction contribution to the Hamiltonian operator can be considered as a perturbation to the standard atomic Hamiltonian used in Eq. (4.1). It can be written as

$$\hat{H}_{SO} = +f(r)\hat{\underline{l}} \cdot \hat{\underline{s}}. \tag{6.6}$$

For a many-electron atom one can write

$$\hat{H}_{SO} = \frac{A(L, S)}{\hbar^2} \hat{\underline{L}} \cdot \hat{\underline{S}}, \tag{6.7}$$

where $\hat{\underline{L}}$ and $\hat{\underline{S}}$ are the total orbital and spin angular momentum operators. $A(L, S)$ is a constant for a given term, that is a given configuration and values of L and S. The spin-orbit coupling is still proportional to $r^{-1}\frac{dV}{dr}$, which becomes r^{-3} for a pure Coulomb potential. This means that A is largest for states with small n.

The energy shifts which result from spin-orbit interaction can be evaluated by considering the expectation value of the spin-orbit Hamiltonian:

$$\Delta E_{SO} = \int \Psi^* \hat{H}_{SO} \Psi d\tau, \tag{6.8}$$

where the integral runs over all space and spin coordinates of the electronic wavefunction Ψ. The wavefunction Ψ for a complex atom is usually not known. However within a non-relativistic framework, the quantum numbers E, L, S, J and M_J are constants of motion and can therefore be taken as known. In particular the angular momentum quantum numbers are obtained as eigenvalues of the following operator equations:

$$\hat{L}^2\Psi = L(L+1)\hbar^2\Psi,$$
$$\hat{S}^2\Psi = S(S+1)\hbar^2\Psi, \qquad (6.9)$$
$$\hat{J}^2\Psi = J(J+1)\hbar^2\Psi.$$

Furthermore, by definition the angular momentum operators obey the relationship

$$\underline{\hat{J}} = \underline{\hat{L}} + \underline{\hat{S}}. \qquad (6.10)$$

This means that

$$\underline{\hat{J}}^2 = (\underline{\hat{L}} + \underline{\hat{S}}) \cdot (\underline{\hat{L}} + \underline{\hat{S}})$$
$$= \underline{\hat{L}}^2 + \underline{\hat{S}}^2 + 2\underline{\hat{L}} \cdot \underline{\hat{S}}$$
$$\therefore \underline{L} \cdot \underline{S} = \frac{1}{2}(\hat{J}^2 - \hat{L}^2 - \hat{S}^2). \qquad (6.11)$$

One can use this operator relationship to evaluate the effect of $\underline{\hat{L}} \cdot \underline{\hat{S}}$ acting on the wavefunction:

$$\underline{\hat{L}} \cdot \underline{\hat{S}}\Psi = \frac{1}{2}(\hat{J}^2 - \hat{L}^2 - \hat{S}^2)\Psi$$
$$= \frac{\hbar^2}{2}[J(J+1) - L(L+1) - S(S+1)]\Psi. \qquad (6.12)$$

The expression for the energy shift due to spin-orbit coupling is therefore

$$\Delta E_{SO} = \frac{A(L,S)}{2}[J(J+1) - L(L+1) - S(S+1)]. \qquad (6.13)$$

This derivation follows the work of Alfred Landé (1888–1976).

Note that Hund's third rule (see Sec. 4.10), means that $A(L,S)$ is positive for an atom with a less than half-full shell, such as sodium, and negative for an atom whose shell is more than half full. Use of Eq. (6.13) is best illustrated by example.

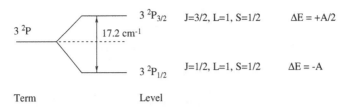

Fig. 6.3. Spin-orbit effects in the $3\,^2P$ term of sodium. ΔE gives the shift of the level relative to the term.

Worked Example 1: The Sodium D Lines.

The transitions $3\,^2P^o_{\frac{1}{2},\frac{3}{2}} \rightarrow 3\,^2S_{\frac{1}{2}}$ in Na I lie in the orange part of the visible spectrum. The 3p ($3\,^2P$) term in sodium is split by spin-orbit interaction, as given in Fig. 6.3.

The sodium D lines are so called because they were so labelled by Fraunhofer in his original solar spectrum (see Fig. 1.1). However, the D line is actually a doublet and the components are usually labelled:

$$D_2 \; 5890\,\text{Å} \; 3p-3s \; 3\,^2P_{\frac{3}{2}}-3\,^2S_{\frac{1}{2}},$$
$$D_1 \; 5896\,\text{Å} \; 3p-3s \; 3\,^2P_{\frac{1}{2}}-3\,^2S_{\frac{1}{2}}.$$

Worked Example 2: The ground state of carbon.

The ground state of carbon has the term ^3P (see Sec. 4.10), and levels given by $J = 0, 1, 2$. The spin-orbit terms are evaluated in Table 6.3. As A is positive for C I, this gives the energy ordering $^3P_2 > {}^3P_1 > {}^3P_0$. Figure 6.4 illustrates the observed splittings.

It can be seen from Fig. 6.4 that the splittings are not exactly in the 2:1 ratio implied by the values in Table 6.3. This is because the treatment given above is highly simplified. There are many other small (magnetic) interactions which need to be considered in a full treatment. However, for low Z atoms and ions, the splitting between the levels approximately follow the intervals given by Eq. (6.13) or what is called the *Landé interval rule*.

Table 6.3. Spin-orbit interaction terms in the ground state ^3P term of the carbon atom.

Level	L	S	J	$\frac{1}{2}[J(J+1) - L(L+1) - S(S+1)]$
$^3P_2$	1	1	2	$+1$
$^3P_1$	1	1	1	-1
$^3P_0$	1	1	0	-2

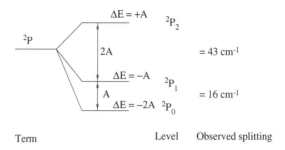

Term   Level   Observed splitting

Fig. 6.4. Spin-orbit effects in the ground ^{3}P term of carbon. ΔE gives the shift of the level relative to the term as predicted by Landé theory; splittings measured in the laboratory are given for comparison.

The value of A increases rapidly with ion charge, approximately as Z_{eff}^4. Thus, for example, the ground state of C I, considered above, has $A \approx 16\,\text{cm}^{-1}$, whereas the isoelectronic O III ion has $A \approx 95\,\text{cm}^{-1}$ in its ^{3}P ground state.

6.3 Fine Structure Transitions

The different transitions between components of spin-orbit states are called 'fine structure transitions'. The important selection rule for these transitions is $\Delta J = 0, \pm 1$ with the exception that $J = 0 \leftrightarrow 0$ transitions are not allowed.

Following this rule, the transition Na $3\,^2D_{\frac{5}{2},\frac{3}{2}} \rightarrow 3\,^2P^o_{\frac{3}{2},\frac{1}{2}}$ has three components because:

$$^2D_{\frac{5}{2}} \rightarrow {}^2P^o_{\frac{3}{2}}, {}^2D_{\frac{3}{2}} \rightarrow {}^2P^o_{\frac{3}{2}} \text{ and } {}^2D_{\frac{3}{2}} \rightarrow {}^2P^o_{\frac{1}{2}} \text{ are all allowed but}$$
$$^2D_{\frac{5}{2}} \rightarrow {}^2P^o_{\frac{1}{2}} \text{ is not}.$$

The three allowed transitions give rise to a triplet structure (see Fig. 6.5).

The strength of the individual fine structure transitions is given by a degeneracy factor times the line strength. Within a non-relativistic treatment, the line strength is determined completely by the terms involved and is therefore the same for all the fine structure transitions. Thus the degeneracy factor gives the relative strength of the fine structure individual components. The actual formula for these degeneracy factors depends on the angular momentum quantum numbers of the upper and lower levels. A general formulation is fairly complicated and only specific answers

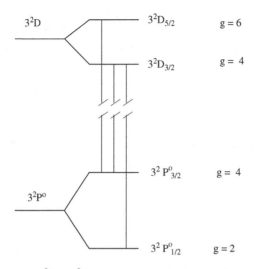

Fig. 6.5. The sodium $3\,^2\mathrm{D}$–$3\,^2\mathrm{P^o}$ triplet, $g = 2J + 1$ gives the statistical weight of each level.

will be quoted here. Tables giving the ratios for all cases can be found in *Allen's Astrophysical Quantities* (see further reading).

Using these tables the relative strength of the triplet $^2\mathrm{D}_{\frac{5}{2},\frac{3}{2}} \rightarrow 3\,^2\mathrm{P^o}_{\frac{3}{2},\frac{1}{2}}$ transition discussed above is

$$^2\mathrm{D}_{\frac{5}{2}} \rightarrow {}^2\mathrm{P^o}_{\frac{3}{2}} : {}^2\mathrm{D}_{\frac{3}{2}} \rightarrow {}^2\mathrm{P^o}_{\frac{3}{2}} : {}^2\mathrm{D}_{\frac{3}{2}} \rightarrow {}^2\mathrm{P^o}_{\frac{1}{2}} \text{ of } 10 : 4 : 2$$

This corresponds to a ratio of 5:2:1. Any spectrum observed in optically-thin conditions should show these ratios. If the spectrum is optically thick then a ratio closer to 1:1:1 will be observed. Thus the intensity ratio between these transitions give direct information on the optical depth of the spectrum.

6.4 Astronomical Sodium Spectra

The sodium resonance line, Na I 3s–3p, is prominent in absorption in the solar spectrum (see Fig. 6.6), and is known as the sodium D spectrum. For an S–P doublet the intrinsic ratio of line intensities is always 2:1. When unsaturated, or optically thin, the strength of the absorption by the doublet that make up this line should be such that the D_1 line is twice as strong as the D_2 line (see Fig. 6.7). This ratio is approximately reproduced in Fig. 6.6 which implies that the lines are optically thin in the Sun. The sodium D lines are also observed in absorption against starlight in the interstellar medium. The D lines are usually very saturated in such spectra.

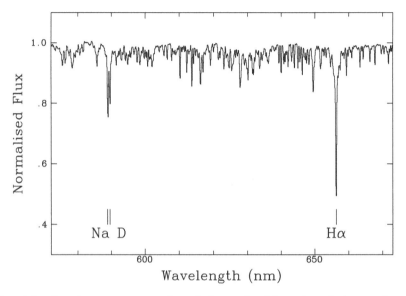

Fig. 6.6. A solar spectrum reflected from the Moon just before a lunar eclipse taken at the University of London Observatory. (S.J. Boyle, private communication.)

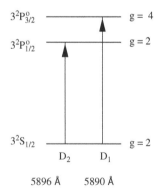

Fig. 6.7. The sodium D lines, $g = 2J + 1$, give the statistical weight of each level.

Figure 6.8 shows the sodium D lines recorded in absorption. This spectrum shows interstellar medium (ISM) sodium ground state atoms absorbing against light from reddened stars. The example shown is strongly saturated as the intensity ratio for the two D lines is almost equal.

Figure 6.8 also shows two features which are called *diffuse interstellar bands* (DIBs). DIBs are ubiquitous ISM absorption features which are present in all reddened lines of sight. They are broader than atomic lines and therefore are almost certainly molecular in origin. There have been many proposed assignments for these features, most of which have been proved to be incorrect and none of which are completely accepted. DIBs have been observed since the 1920's and the long-running failure to resolve the DIBs problem has been described by Patrick Thaddeus as 'the scandal of modern astronomy'.

Besides the sodium D lines, the Na I 3s – 4p doublet at 3302 Å can also be observed in absorption in the ISM. These transitions are much weaker

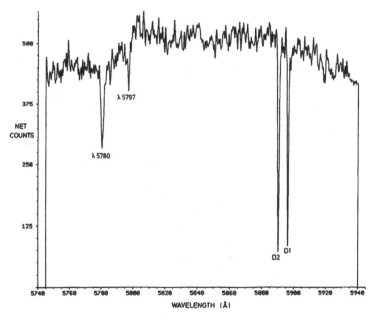

Fig. 6.8. Interstellar absorption by the sodium D lines recorded towards HD 97950 which is the central object in compact galactic cluster NGC 3603, recorded using the Anglo-Australian Telescope. Note the two unassigned absorption features which are due to diffuse interstellar bands. [Reproduced from W.B. Somerville and J.C. Blades, *Mon. Not. R. Astron. Soc.* **192**, 719 (1980).]

than the D lines, $A(3s-4p) = 2.8 \times 10^6 \text{ s}^{-1}$, compared to $A(3s-3p) = 6.2 \times 10^7 \text{ s}^{-1}$. This means that the 3s–4p doublet can be used when D lines are saturated. However they lie in a difficult spectral region where there are many telluric (i.e. atmospheric) features due to ozone.

Higher transitions of sodium are often also observed and provide important spectral markers in the atmosphere of cool stars. Figure 6.9 shows the spectrum of an L-subdwarf star, a cool star with a mass about 8% of our Sun. Absorption features due to Na I, K I and Rb I are clearly visible in the spectrum along with much more complicated molecular features.

All the alkali metal features appear to be doublets. In fact the Na I absorption features arise from a triplet 3p–3d transition:

$$8183.26 \text{ Å } 3\,^2P^o_{\frac{1}{2}} - 3\,^2D_{\frac{3}{2}} \, ;$$

$$8194.79 \text{ Å } 3\,^2P^o_{\frac{3}{2}} - 3\,^2D_{\frac{3}{2}} \, ;$$

$$8194.82 \text{ Å } 3\,^2P^o_{\frac{3}{2}} - 3\,^2D_{\frac{5}{2}} \, .$$

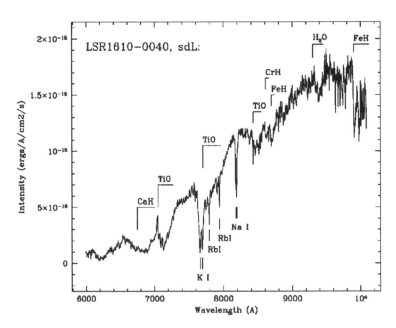

Fig. 6.9. Spectrum of the L-subdwarf star LSR 1610–0400 showing clear spectral features due to Na I, K I and Rb I as well as diatomic molecules CaH, TiO, CrH and FeH, and water. [Reproduced from S. Lépine, R.M. Rich and M.M. Shara, *Astrophys. J.* **591**, L49 (2003).]

This appears as a doublet since the splitting in the upper 3 ^{2}D term is too small to be resolved.

The K I and Rb I transitions both belong to the resonance line which is a doublet in each case:

$$\text{K I } 7664.91\,\text{Å } 4\,^2S_{\frac{1}{2}} - 4\,^2P^o_{\frac{3}{2}}\,;$$

$$\text{K I } 7698.97\,\text{Å } 4\,^2S_{\frac{1}{2}} - 4\,^2P^o_{\frac{1}{2}}\,.$$

$$\text{Rb I } 7800.27\,\text{Å } 5\,^2S_{\frac{1}{2}} - 4\,^2P^o_{\frac{3}{2}}\,;$$

$$\text{Rb I } 7947.60\,\text{Å } 5\,^2S_{\frac{1}{2}} - 4\,^2P^o_{\frac{1}{2}}\,.$$

The significantly larger splitting in the Rb I resonance line is due to the larger spin-orbit effects in this heavy atom. Indeed the transitions of all three species give a nice illustration of the variation of spin-orbit effects with principal quantum number n, and atomic number Z.

6.5 Other Alkali Metal-Like Spectra

There are a number of astronomically important ions whose ground states consist of a single s electron outside a closed shell. These ions have spectra similar to those of the alkali metals. Examples include:

Ca II or potassium-like calcium

The transitions

$$\text{H line: } 4\,^2S_{\frac{1}{2}} - 4\,^2P^o_{\frac{1}{2}} \text{ at } 3968.47\,\text{Å},$$

$$\text{K line: } 4\,^2S_{\frac{1}{2}} - 4\,^2P^o_{\frac{3}{2}} \text{ at } 3933.66\,\text{Å},$$

are seen in the solar spectrum. Indeed the labels 'H' and 'K' are due to Fraunhofer. These lines absorb strongly in cool stars (see Fig. 6.10). The H line is often blended with the hydrogen Balmer line Hϵ at 3970.07 Å. High-resolution studies of the structure of K line absorption profiles observed in the Sun can give detailed information on the vertical distribution of Ca$^+$ in the solar chromosphere.

Calcium has an ionisation potential of 6.1 eV, significantly lower than hydrogen. This means that Ca$^+$ is found in the ISM, where it can be observed unblended with Hϵ, since Hϵ involves a transition between two excited states, where such excited state lines are not present in the cold ISM. The unsaturated intensity ratio of the K to H line in absorption is 2 to 1, as found for the Na D lines, as this is a general property of S–P doublets.

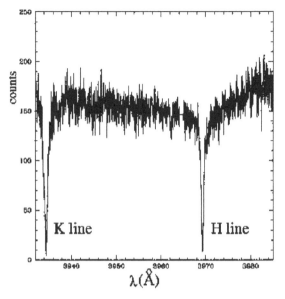

Fig. 6.10.   High-resolution spectrum of the cool, hydrogen-rich white dwarf star WD 1633+4333 showing Ca II H and K lines, recorded using the Keck Telescope. [Adapted from B. Zuckerman *et al.*, *Astrophys. J.* **596**, 477 (2003).]

Ca II also has a strong triplet transition in the red, sometimes referred to as the Ca T lines. These lines are:

$$8498.0\,\text{Å}\ 4\,^2P^o_{\frac{3}{2}} - 3\,^2D_{\frac{3}{2}},$$

$$8542.1\,\text{Å}\ 4\,^2P^o_{\frac{3}{2}} - 3\,^2D_{\frac{5}{2}},$$

$$8662.1\,\text{Å}\ 4\,^2P^o_{\frac{1}{2}} - 3\,^2D_{\frac{3}{2}}.$$

Absorptions due to the Ca T lines are clearly visible in the atmosphere of cool stars where they are used as an important diagnostic test of metallicity, the proportion of atomic species heavier than helium (see Fig. 6.11).

Mg II *or sodium-like magnesium*

The transitions:

$$2802.7\,\text{Å}\ 3\,^2S_{\frac{1}{2}} - 3\,^2P^o_{\frac{1}{2}},$$

$$2795.5\,\text{Å}\ 3\,^2S_{\frac{1}{2}} - 3\,^2P^o_{\frac{3}{2}},$$

lie in the ultraviolet and so have to be observed by satellite. These lines can be routinely monitored by the Hubble Space Telescope (see Fig. 6.12).

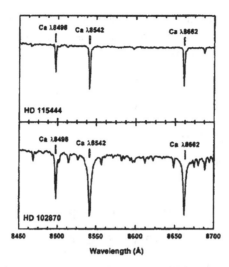

Fig. 6.11. Ca II triplet red absorption spectra in stars HD 115444 (*upper*) and HD 102870 (*lower*) recorded at the Observatoire de Haute Provence. HD 102870 is an F-type star and has a significantly higher abundance of metals than K-star HD 115444. These give rise to the other weaker features. [Reproduced from T.P. Idiart, F. Thévenin and J.A. de Freitas Pachieco, *Astron. J.* **113**, 1066 (1997).]

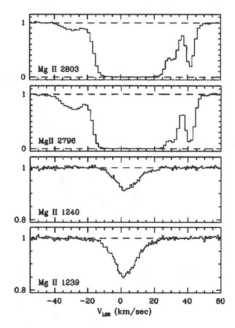

Fig. 6.12. Ultraviolet interstellar Mg II absorption-line measurements for the sight line towards O-star μ Columbae, obtained with the Hubble Space Telescope. [Adapted from J.C. Howk, B.D. Savage and D. Fabian, *Astrophys. J.* **525**, 253 (1999).]

Mg II lines are observed as strong absorptions in stellar atmosphere and in the ISM, and as emissions from circumstellar shells. If these lines are saturated, the weaker transitions

$$1240.4\,\text{Å}\ 3\ ^2S_{\frac{1}{2}} - 4\ ^2P^o_{\frac{1}{2}},$$
$$1239.9\,\text{Å}\ 3\ ^2S_{\frac{1}{2}} - 4\ ^2P_{\frac{3}{2}},$$

can be monitored instead.

C IV *or lithium-like carbon*

The transitions

$$1550.8\,\text{Å}\ 2\ ^2S_{\frac{1}{2}} - 2\ ^2P^o_{\frac{1}{2}},$$
$$1548.2\,\text{Å}\ 2\ ^2S_{\frac{1}{2}} - 2\ ^2P^o_{\frac{3}{2}},$$

were also extensively studied by the International Ultraviolet Explorer (IUE) (see Fig. 6.13). These lines are very prominent in emission from circumstellar shells and quasars, and can also be seen in absorption.

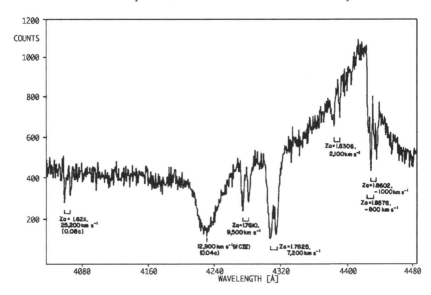

Fig. 6.13. C IV emission and absorption lines towards Q 00135.01–4001 recorded using the International Ultraviolet Explorer satellite. This figure illustrates three types of absorption lines found in quasi-stellar objects (QSOs). Type A, the broad trough about λ4230; Type B, the sharp pair of lines at $z \sim 1.86$; and four Type C doublets at $Z = 1.62$, 1.76, 1.78 and 1.83. [Adapted from R.J. Weymann, R.F. Carswell and M.G. Smith, *Ann. Rev. Astron. Astrophys.* **19**, 41 (1981).]

All the above lines show the characteristic fine structure patterns of the alkali metals. This fine structure behaviour has itself been the subject of extensive recent studies to try and address the question of whether physical constants remain constant with time. The fine structure splittings all depend on a dimensionless physical constant called the 'fine structure constant', whose value, in atomic units, is simply the inverse of the speed of light. By measuring fine structure splittings for a large range of redshifts, it is possible to determine the value of the fine structure constant at different epochs. The present evidence point towards a gradual change in this constant with time suggesting that the physical constants in our Universe did not always have today's values. However this result awaits confirmation.

Problems

6.1 By considering the levels that arise from the following configurations and dipole selection rules, determine what spectral lines would be produced in the following transitions:

(a) Na I 3p – 4d,
(b) Na I 3d – 5f,
(c) Na I 4s – 4d,
(d) K I 4s – 4p.

Comment on the possible observability of these transitions in different astronomical locations.

6.2 A series of transitions in atomic potassium consists of emissions from several np levels to the 4s atomic ground state. Transitions are observed at $12985 \, \text{cm}^{-1}$ when $n = 4$; $24701 \, \text{cm}^{-1}$ when $n = 5$; $28999 \, \text{cm}^{-1}$ when $n = 6$. The series limit, which can be assumed to be the same as the ionisation potential of the 4s level, is at $35010 \, \text{cm}^{-1}$. Calculate the quantum defects of the 4p, 5p and 6p levels. Hence estimate the wavenumber of the corresponding $n = 7$ transition.

6.3 The potassium transitions shown in Fig. 6.9 are a doublet linking the ground state to the $4\,^2\text{P}^{\text{o}}_{\frac{3}{2}}$ and $4\,^2\text{P}^{\text{o}}_{\frac{1}{2}}$ levels. The two lines lie at 764.494 nm and 769.901 nm, respectively. Use this information to calculate the constant of proportionality, $A(L, S)$, in the expression for the spin-orbit interaction energy, Eq. (6.13).

6.4 Observations are made of atomic sodium emitting from its $1s^2 2s^2 2p^6 4d$ configuration to its $1s^2 2s^2 2p^6 3p$ configuration. At high resolution, three transitions are observed. Label these transitions using spectroscopic notation. A lower resolution survey spectrum only resolves the Na 4d – 3p emission as a single line. What other Na emission lines must be present in the spectrum? Suggest an astronomical location where the Na 4d – 3p transitions might be observed. An astronomical observation resolves all three transitions in absorption but finds them to be of approximately equal intensity. Can one determine the column density of Na atoms?

6.5 Configurations of triply ionised carbon, C^{3+}, can be written $1s^2nl^1$. In terms of R_∞ and the quantum defect, give an expression for the energy levels of this system. Briefly explain the physical significance of the quantum defect and how it depends on n and l.

The ionisation energy of C^{3+} is $520178\,cm^{-1}$ in its $1s^2 2s^1$ ground state and is $217329\,cm^{-1}$ in the $1s^2 3s^1$ state. Estimate the ionisation energy of the $1s^2 4s^1$ state of C^{3+}.

6.6 The $1s^2 2s^2 2p^3$ ground state configuration of O^+ (oxygen-like nitrogen) leads to three terms: $^4S^o$, $^2P^o$ and $^2D^o$.

(a) What energy order would you expect the terms to be in?
(b) Give the levels for each term in full spectroscopic notation.
(c) Assuming a spin-orbit splitting constant $A' = 30\,cm^{-1}$, sketch the pattern of level splitting for each of the doublet terms.

6.7 In spectroscopic notation an atom has a level which is designated $^3F^o_2$. Explain the meaning of this symbol and give values for the angular momenta it represents. What other levels arise from the same term? Suggest a configuration of atomic carbon that could give rise to this level.

The $^3F^o_2$ level can emit to levels $^3P_2$, $^3P^o_2$, $^1D_3$ and $^3D_3$. Assuming electric dipole selection rules, order these transitions by their probable strength, giving your reasons.

CHAPTER SEVEN

SPECTRA OF NEBULAE

'Apparelled in celestial light'

– William Wordsworth, *Intimations of Immortality* (1807)

Nebulae often have particularly rich spectra (see Figs. 7.1 and 7.2 for examples). The spectra contain a wealth of information and can allow pictures to be built up of the different regions involved. The spectra of nebulae largely involve atomic emissions. These emissions are driven by a number of different physical mechanisms which are responsible for creating atoms and ions in excited states which allow them to emit. Each mechanism has a different spectral signature. The primary mechanisms are:

Electron collisions: electron collisions largely populate low-lying excited states. Excitations are essentially thermal in nature and depend on the electron temperature T_e.

Recombination: the cascade of emissions following the recombination of an electron with an ion leads to emissions from more highly-excited states than electron collisions.

Optical pumping (or resonance-fluorescence): a special mechanism which depends on the detailed physics of the system being observed. As discussed below, optical pumping is characterised by rather specific line emissions.

7.1 Nebulium

In 1918, extensive studies of the emission spectra of nebulae found a series of lines which had not been observed in the laboratory. Particularly strong

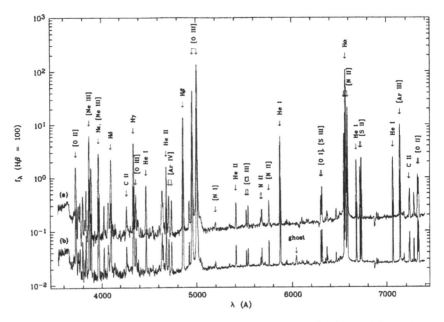

Fig. 7.1. Optical spectra of NGC 6153 from 3540 to 7400 Å, obtained for a deep (10-minute) exposure using the ESO 1.52 m telescope in Chile. The two spectra plotted are (a) obtained by uniformly scanning the long-slit across the entire nebula, and (b) taken with a fixed slit centred on the central star. [Reproduced from X.-W. Liu *et al.*, *Mon. Not. R. Astron. Soc.* **312**, 585 (2000).]

were features at 4959 Å and 5007 Å. For a long time this pair could not be identified and, inspired by Lockyer's success with helium, these lines were attributed to a new element, 'nebulium'.

Ten years after their original observation, Ira Bowen (1898–1973) found the true explanation. Bowen realised that in the diffuse conditions found in nebulae, atoms and ions could survive a long time without undergoing collisions. Indeed, under typical nebula conditions the mean time between collisions is in the range 10–10000 s. This means that there is sufficient time for excited, metastable states to decay via weak, forbidden line emissions. These lines could not be observed in the laboratory where it was not possible to produce collision-free conditions over this long timeframe.

Bowen identified the doublet as transitions within the $1s^2 2s^2 2p^2$ ground state configuration of O^{2+}:

$$5006.84 \text{ Å } [O\,\textsc{iii}] \; 1s^2 2s^2 2p^2 \; {}^1D_2 - {}^3P_2,$$
$$4958.91 \text{ Å } [O\,\textsc{iii}] \; 1s^2 2s^2 2p^2 \; {}^1D_2 - {}^3P_1.$$

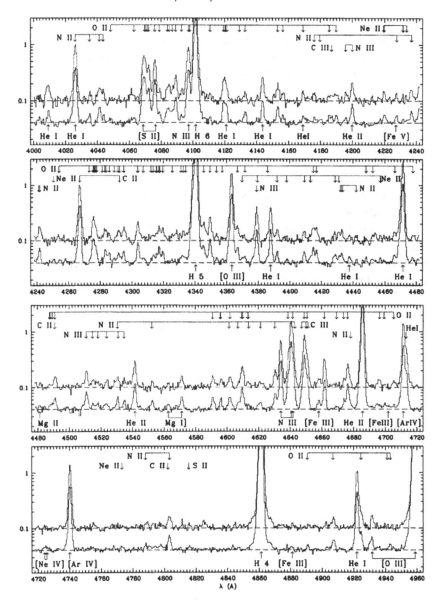

Fig. 7.2. Continuum-subtracted spectra of planetary nebula NGC 6153 from 4000 to 4960 Å obtained using the ESO 1.52 m telescope in Chile. The spectra show the rich recombination-line spectra from C, N, O and Ne ions. The upper spectrum was obtained by uniformly scanning the entire nebular surface using a narrow long-slit, and the lower one was obtained with a fixed slit centred on the central star. The spectra were normalised such that the flux of Hβ = 100. [Reproduced from X.-W. Liu *et al.*, *Mon. Not. R. Astron. Soc.* **312**, 585 (2000).]

Such transitions are strongly forbidden for electric dipoles by the Laporte rule. They only occur as weak magnetic dipole transitions. Even for magnetic dipole lines these transitions are weak as they are also spin-changing, intercombination transitions. They have Einstein A coefficients of $1.8 \times 10^{-10}\,\mathrm{s}^{-1}$ and $6.2 \times 10^{-11}\,\mathrm{s}^{-1}$, respectively. The line

$$4931.23\,\text{Å}\ [\text{O\,III}]\ 1s^2 2s^2 2p^2\ {}^1D_2 - {}^3P_0$$

is even weaker since this is magnetic dipole-forbidden and occurs as an electric quadrupole transition with $A = 2.4 \times 10^{-14}\,\mathrm{s}^{-1}$.

Other 'nebulium' lines turned out to be due to [O\,II] and [N\,II] forbidden transitions.

The forbidden spectrum of the O^{2+} ion, [O\,III], are often strong in nebulae. There are a number of [O\,III] transitions within the ground state configuration of O^{2+} $1s^2 2s^2 2p^2$ (see Fig. 7.3). As shown in Sec. 4.10, $2p^2$ has terms ${}^3P$, ${}^1D$, ${}^1S$ in that energy order. All transitions between these terms

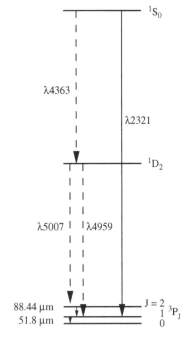

Fig. 7.3. [O\,III] transitions within the O^{2+} ground state configuration $1s^2 2s^2 2p^2$. All transitions are electric dipole-forbidden by the Laporte rule.

are strongly dipole-forbidden by the Laporte rule. The observed transitions are therefore electric quadrupole or magnetic dipole ones.

The excited terms associated with the ground state configuration are populated by electron collisions. The ^{1}S term is higher in energy and therefore requires collisions with more energetic electrons to be populated. This line decays via an electric quadrupole transition emitting green light familiar from the earth's aurora

$$4363.2 \, \text{Å [O III]} \, 1s^2 2s^2 2p^2 \, {}^1S_0 - {}^1D_2 \, .$$

The emission rate from ^{1}D relative that to ^{1}S depends directly on the electron temperature T_e. To interpret these data it is necessary to know from laboratory studies not only the relevant transition probabilities for the observed lines but also the branching ratios, which give the proportion of ions which decay via these transitions as opposed to the alternative decay routes which also exist. With this information the ratio of the unsaturated, observed line strengths can be used to give a measure of T_e.

[N II] is also C-like and therefore has a similar structure to [O III]. [N II] can also be used to measure T_e. In the [O III] and [N II] examples, the upper states have different energies but similar lifetimes. Together they provide a useful thermometer. Other transitions can provide a measure of electron density.

A good example of this is the C III doublet at 1907 and 1909 Å. These two transitions both involve the same configurations but while the stronger transition,

$$1908.73 \, \text{Å C III]} \, 2s^2 \, {}^1S_0 - 2s2p \, {}^3P_1^o$$

is an intercombination line with Einstein A coefficient of $114 \, \text{s}^{-1}$, the other weaker line,

$$1906.68 \, \text{Å [C III]} \, 2s^2 \, {}^1S_0 - 2s2p \, {}^3P_2^o$$

is completely electric dipole-forbidden as $\Delta J = 2$. This transition occurs as a very weak magnetic quadrupole transition with $A = 0.005 \, \text{s}^{-1}$. The huge difference in lifetime for the two transitions is reflected in their critical density and hence the relative strength of the two lines can be used as a sensitive probe of electron density.

7.2   The Bowen Mechanism

In Planetary Nebulae (PN), many emission lines of O III can be observed at
visible wavelengths. Some of these transitions are between excited states
of O^{2+} which are far too high in energy to be populated by collisions. In
particular there is a whole series of lines in the 3100–3800 Å region (see
Fig. 7.5). These correspond to electric dipole allowed transitions of O III
but between high-lying states.

The 3s – 3p lines which lie in the 3100–3800 Å region follow as a result
of a cascade from 3p – 3d transitions. There therefore has to be some non-
thermal mechanism whereby the $1s^2 2s^2 2p3d$ configuration is preferen-
tially populated. Bowen realised that the $^3P^o_2$ level of this configuration
can be populated through an accidental resonance. The He II Lyα line is at
303.78 Å; the transition

$$303.80 \text{ Å O III } 2p^2 \; {}^3P_2 - 2p3d \; {}^3P^o_2$$

(note that the closed shell $1s^2 2s^2$ is assumed) lies at almost the same wavelength as He II Lyα. The small difference is covered by thermal (Doppler)
shifts within nebula.

Helium can be doubly ionised in hot nebulae and hence the recombination spectrum of He II leads to a plentiful supply of Lyα photons. These
then excite O^{2+}, specifically its 2p3d $^3P^o_2$ level. This level then decays by
emission. This mechanism is depicted schematically in Fig. 7.4.

The Bowen mechanism is an example of a physical process known
as *optical pumping* or *resonance-fluorescence*. Although such processes rely
on an accidental coincidence in level spacing between different species, a
number of such occurrences are now known to be astronomically important. For example hydrogen Lyβ photons at 1025.72 Å can pump the
transition:

$$1025.77 \text{ Å O I } 2p^4 \; {}^3P_2 - 2p^3({}^4S^o)3d \; {}^3D^o \, .$$

In this case, unlike O III, the fine structure splitting in the upper level is so
small that all three $^3D^o$ levels are populated.

In an ordinary H II region, there are fewer high-energy, ultraviolet
photons so there are very little He^{2+}. Under these circumstances, O^{2+}
ions are still present and the forbidden [O III] lines are seen. However the
Bowen mechanism cannot operate and the allowed lines are not observed.

O^{2+} transitions can also be observed in the infrared, see Fig. 7.6. In
this figure note also the Balmer transitions H12α and H13α. The O^{2+} lines

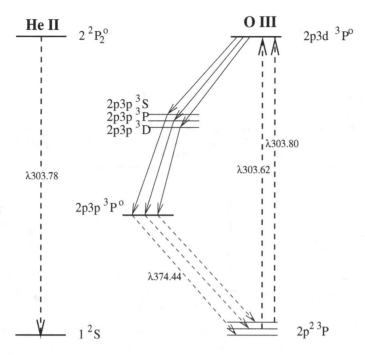

Fig. 7.4. Schematic partial energy level diagram of O III and He II showing the coincidence of He II Lyα and O III $2p^2$ $^3P_2$ – $2p3d$ $^3P^o_2$ transitions. The Bowen resonance-fluorescence lines in the visible and ultraviolet are indicated by the solid lines.

are two forbidden fine structure transitions within the $2p^2$ ground state configuration:

$$51.8 \ \mu m \ [O III] \ 2p^2 \ ^3P_1 - {}^3P_2 \ ;$$
$$88.4 \ \mu m \ [O III] \ 2p^2 \ ^3P_0 - {}^3P_1 \ .$$

These are weak magnetic dipole transitions. These infrared transitions could not be seen from the ground but were observed by the Infrared Space Observatory (ISO).

It should be noted that the spectrum of NGC 7027 (Fig. 7.6) contains transitions assigned to various molecules and molecular ions; the spectra of these species are discussed in Chapter 10. The planetary nebula NGC 7027 has a particularly rich chemistry and the spectrum contains a number of lines whose identity is either unknown or uncertain.

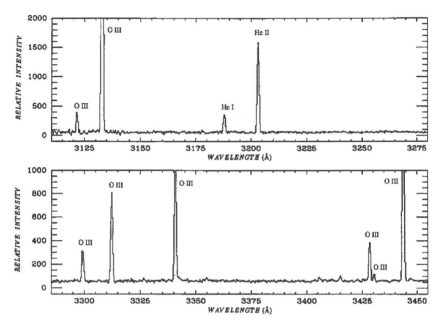

Fig. 7.5. O III emission lines excited by the Bowen mechanism from planetary nebula NGC 3242, recorded using the Isaac Newton Telescope at La Palma. [Adapted from X.-W. Liu and J. Danziger, *Mon. Not. R. Astron. Soc.* **261**, 465 (1993).]

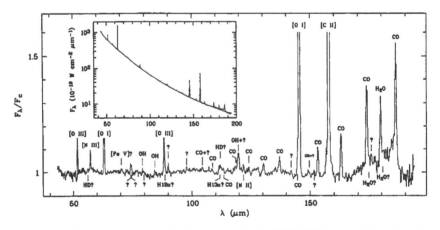

Fig. 7.6. Infrared spectrum of planetary nebula NGC 7027 recorded using the Infrared Space Observatory (ISO). The spectrum is observed on a rapidly varying background, as shown in the insert, and the detailed line spectrum is obtained after subtracting a polynomial fit to the continuum part of the spectrum. [Reproduced from X.-W. Liu *et al.*, *Astron. Astrophys.* **315**, L257 (1996).]

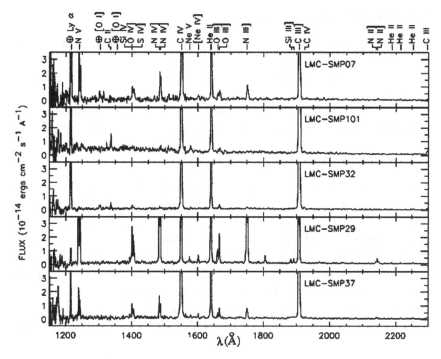

Fig. 7.7. Ultraviolet spectra of planetary nebulae in the Large Magellanic Cloud recorded using the Hubble Space Telescope. The Lyα emissions are geocoronal in origin. [Adapted from E. Vassiliadis *et al.*, *Astrophys. J. Suppl. Ser.* **114**, 237 (1998).]

O^{2+} transitions are also observed in the ultraviolet (see Fig. 7.7). The O III] doublet near 1663 Å are a pair of intercombination lines, as shown by the single bracket, meaning that the electric dipole selection rules are all satisfied except for spin conservation, i.e. $\Delta S \neq 0$. These transitions link the lowest quintet state to the ground state, which is a triplet:

$$1660.81 \text{ Å O III] } 1s^2 2s^2 2p^2 \ ^3P_1 - 1s^2 2s 2p^3 \ ^5S_2^o \,.$$

$$1666.15 \text{ Å O III] } 1s^2 2s^2 2p^2 \ ^3P_2 - 1s^2 2s 2p^3 \ ^5S_2^o \,.$$

The upper level is just a little higher in energy than the ground configuration ^1S. It can therefore be populated by collisions with energetic electrons. Other terms with the configuration $2s2p^3$ are $^3S^o$, $^3D^o$, $^3P^o$, $^1P^o$ and $^1D^o$. These all have higher energy and so the lines linking them to the ground state are at still shorter wavelengths. These transitions can be used to monitor still higher electron temperatures.

7.3 Two Valence Electrons

The previous chapter considered the spectra of alkali metals and alkali metal-like ions which all have a single optically-active electron. The presence of two outer shell electrons introduces important new physical processes.

The calcium atom will be used to illustrate these processes. The ground state of calcium has two 4s electrons outside a closed shell:

$$1s^2 2s^2 2p^6 3s^2 3p^6 4s^2 .$$

Astronomical spectra of Ca I can be understood entirely in terms of these two outer electrons, which are called *valence* or *optically-active* electrons. One can therefore safely assume that the other electrons are inactive and remain frozen in their ground state orbitals.

Excitation of one 4s electron gives the configuration 4snl which can have spin $S = 0$ or 1, except for the 4s^2 configuration itself which gives rise to $S = 0$ only. The orbital angular momentum possessed by these states is $L = l$. The resulting energy level diagram for the 4snl states of Ca is given by Fig. 7.8. For these states, the ionisation limit, represented

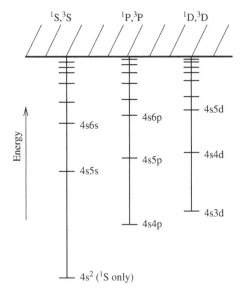

Fig. 7.8. Main branch energy level diagram of calcium. The diagram gives 4snl configurations of the outer two electrons. These converge to Ca$^+$ in its 4s ground state plus a continuum electron.

by $n \rightarrow \infty$, is Ca$^+$ in its 4s configuration, which is the ground state of the ion. The diagram represents what is called the 'main branch' of the Ca I spectrum. However it is not sufficient to give all the observed transitions of Ca I.

To explain how the other transitions arise it is necessary to consider bound states of the calcium atom which have both outer electrons excited. For example, consider the case where one electron is in the 3d orbital and the other electron jumps, giving a configuration 3dnl. This gives rise to many new electronic states which involve excitation of two electrons. These are depicted in Fig. 7.9. The 3dnl states go to an ionisation limit as $n \rightarrow \infty$ of Ca$^+$ in its first excited 3d configuration. A complete understanding of the spectrum of Ca I also requires 4pnl configurations going to the Ca$^+$ 4p configuration at their ionisation limit.

Structure of this sort is important in spectra of nearly all complex atoms. An exception is helium for which all states with two excited electrons are unbound.

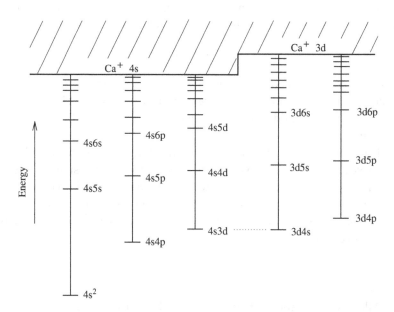

Fig. 7.9. Energy level diagram of calcium. The diagram gives 4snl and 3dnl configurations of the outer two electrons. These converge to Ca$^+$ in its 4s ground state and Ca$^+$ in its 3d first excited state respectively, plus a continuum electron.

7.4   Autoionisation and Recombination

As illustrated by the various series for calcium, each series of energy levels converges towards its relevant continuum level. That is, at the ionisation limit for each series, there remains an ion in some state and an electron in the continuum. However, series involving two-electron excited states, i.e. ones which are not on the main branch, result in ion states which are also excited states. As illustrated by Fig. 7.9, the higher states of such series lie in the continuum of the ground state of the ion. In other words, there are states with two excited electrons which lie above the first ionisation threshold. Such states have the same energy as a state of the ion plus a free electron. Free electrons lie in the continuum since their energy levels are not quantised.

When there is resonance between two states with the same energy, the system goes rapidly from one state to the other and back. 'Bound' states in the continuum are therefore called 'resonances' since they (always) lie at the same energy as an unbound, continuum state. The states of radioactive nuclei are also resonances; these states of the nucleus are ones which spontaneously decay with emissions of radioactive particles or rays. Similarly, electronic resonances can change spontaneously from being a doubly-excited state to an ionised state and an free electron:

$$A^{**} \rightarrow A^+ + e^- . \tag{7.1}$$

This process is called *autoionisation* and is illustrated schematically by Fig. 7.10. It is a radiationless transition in which one electron jumps to a lower level and the other electron escapes to the continuum simultaneously.

Transition probabilities between bound states and resonances obey the same selection rules as transitions between truly bound states. However resonances do not have a precise energy but have some width or spread of energies, ΔE. This is a consequence of Heisenberg's Uncertainty Principle: as the resonance has a finite lifetime, τ, it cannot have a definite energy. The Uncertainty Principle gives

$$\Delta E \cdot \tau \geq \frac{\hbar}{2} . \tag{7.2}$$

Allowed transitions between bound states are generally much stronger than bound-free transitions. This leads to the resonant enhancement of autoionisation (see Fig. 7.11). As allowed transitions to resonances

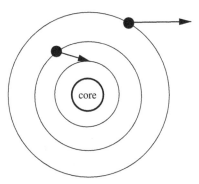

Fig. 7.10. A schematic representation of autoionisation: a radiationless process whereby a system with two excited electrons loses one to the continuum while the other simultaneously jumps to a lower energy level.

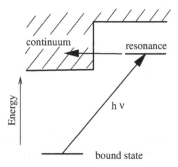

Fig. 7.11. Resonance-enhanced ionisation involves an allowed transition from a bound to a resonance state followed by a radiationless transition to the continuum.

are also stronger, the presence of a suitable resonance can increase significantly the overall ionisation rate in a given radiation field. See Fig. 7.12 for example.

Resonances therefore play an important role in photoionisation. Astronomically they play an even more important role in recombination. As discussed in Sec. 3.9, recombination is an important source of emission in regions where some or all of the atoms are ionised. For complex atoms there are two possible recombination mechanisms:

(1) *'Direct' radiative recombination,* as discussed for H-like atoms in Sec. 3.9.
(2) *Dielectronic recombination*: this can be thought of as the reverse of resonant photoionisation. A continuum electron with the correct energy

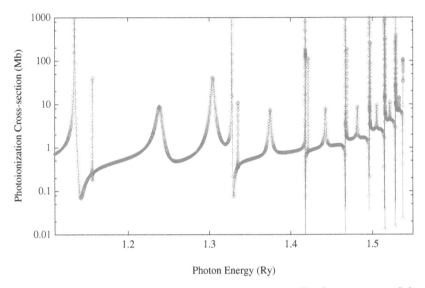

Fig. 7.12.   Photoionisation cross section, in units of $10^{-22}\,\text{m}^2$, for the $2s2p^2\ ^2D$ excited term of C^+ as a function of photon energy calculated by A.R. Davey, P.J. Storey and R. Kisielius, *Astron. Astrophys. Suppl. Ser.* **142**, 85 (2000). The rate of this process is dominated by the near-threshold resonance. (P.J. Storey, private communication.)

forms a resonance by undergoing a radiationless transition. This resonance can then decay by autoionisation or it can become a truly bound state by the emission of a photon.

The process radiative recombination can thus be written as

$$A^+ + e^- \rightarrow A^{**} \rightarrow A^* + h\nu$$
$$\rightarrow A^+ + e^- . \tag{7.3}$$

As autoionisation is a rapid process, only dipole-allowed radiative transitions can compete with it. As the resonance is a two-electron excited state, dielectronic recombination often results in different transitions, i.e. photons with different wavelengths, than direct radiative recombination which only goes via the main branch (see Fig. 7.13).

Dielectronic recombination relies on the presence of resonances which are low-lying enough to be accessible by thermal electrons. Whether such resonances exist is a property of the physics of the particular ion in question. Dielectronic recombination is particularly important in some

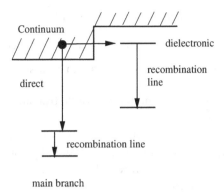

Fig. 7.13. Dielectronic and direct (or radiative) recombination.

cases, such as C^+. C^+ has a very large and low-lying resonance (see Fig. 7.12), formed by the temporary capture of an electron by C^{2+}. C II recombination spectra are completely dominated by lines from dielectronic recombination.

Snapshot spectra, which only survey the strongest lines, show nebula emissions dominated by hydrogen recombination lines plus a few atomic (ion) transitions. However deep spectra which probe weak transitions can show a huge number of lines. See Fig. 7.2 for example. These lines generally arise from atomic recombination spectra. But why are there so many of them? This question is best answered by considering an example.

The strong C II line at 4267 Å in Fig. 7.2 is due to the recombination spectrum of C II. The C^{2+} core is a closed shell with the configuration $1s^2 2s^2 (^1S)$. The resulting recombination lines are due to transitions of the form

$$\text{C II } 1s^2 2s^2 (^1S) \, nl - 1s^2 2s^2 (^1S) \, n'l'.$$

Thus the line at 4267 Å is a 3d – 4f transition. This transition is seen as only a single line as the (triplet) fine structure is not resolved.

Figure 7.2 shows a large number of weaker O II recombination lines. These lines also arise from the 3d ← 4f recombination transition but occur at many wavelengths. The ground state of O^{2+}, which gives rise to the main branch recombination lines, is an open shell: $1s^2 2s^2 2p^2 (^3P)$. When recombined with an extra electron, the open core can couple with the extra electron to form a number of different terms as both the core and the extra electron have spin and orbital angular momenta. Thus

> ^{3}P plus 4f ^{2}F gives 6 separate terms and 18 distinct levels;
>
> ^{3}P plus 3d ^{2}D also gives 6 terms and 16 distinct levels.

There are therefore a large number of possible transitions between the different terms and/or levels. Which transition is the strongest? Since for a given atom or ion all 3d $\leftarrow$ 4f transitions have similar line strengths, the relative strength of the transitions are given by considering statistical weights. For the O II example above, the strongest emission lines which come from the 4f configuration arise from the ^{4}G^o term which has a statistical weight of $g = (2L + 1) \times (2S + 1) = 9 \times 4 = 36$. In contrast the ^{2}D^o term has a statistical weight of only $g = 5 \times 2 = 10$.

Problems

7.1 By considering the levels which arise from the following configurations and dipole selection rules, determine what spectral lines would be produced in the following transitions:

(a) Ca I 4s^2 – 4s4p,

(b) Ca I 4s4f – 4s5d?

Comment on the possible observability of these transitions in different astronomical locations.

7.2 Give the ground state configuration of the N$^+$ ion and explain which of its terms you would expect to have the lowest energy.

A series of N I recombination transitions can occur as a result of radiative recombination of ground state N$^+$ ions and an electron. Many optical 5f – 4d transitions are observed. Explain why so many transitions are seen. Neglecting fine structure effects, how many allowed transitions would you expect? Which one is likely to be the strongest?

7.3 Use Figs. 7.3 and 7.4 to construct a partial energy level diagram for the O^{2+} ion, with energies in eV.

What is the statistical weight of the atomic level $^{2S+1}L_J$? Give the statistical weight for the O^{2+} ion states in your diagram.

Assuming the population of the 2p^2 ^{3}P$_0$ ground state to be 1, use the Boltzmann distribution for thermal populations, and a temperature of $T = 10000$ K, to give the occupation of the other O^{2+} levels in your diagram relative to the ground state.

Discuss the significance of this result in relation to the observed O III spectrum in a nebula with $T_e = 10000\,\mathrm{K}$.

The following will be useful:

$1\,\mathrm{cm}^{-1} = 1.23985 \times 10^{-4}\,\mathrm{eV}$,

Boltzmann's constant, $k = 8.62 \times 10^{-5}\,\mathrm{eV \cdot K^{-1}}$.

CHAPTER EIGHT

X-RAY SPECTRA

X-rays are high-energy photons which have shorter wavelengths than those in the ultraviolet. X-rays can be produced as a result of transitions involving an inner shell in heavy atoms and their name was initially associated with this particular process. Indeed X-rays produced in this fashion in X-ray tubes as a result of some heavy metal being bombarded by high-energy electrons remain important for routine medical inspections of bones inside living people.

Photons with even higher energies than X-rays are called γ-rays. Like X-rays, γ-rays were originally associated only with the process that produced them. In the case of γ-rays this was the radioactive emissions from nuclei. However the terms X-ray and γ-ray are now just used to label parts of the electromagnetic spectrum. Indeed astronomically neither of the two original processes are major sources of high-energy photons.

The previous chapters have considered spectra associated with the movements of outer electrons which result in a so-called 'optical' spectrum. For an atom, with (fairly) high Z, transitions which involve movements of inner shell electrons occur at much shorter wavelengths. In principle, so long as only one electron jumps, there is no difference between the two cases and much of the theory already outlined still applies. In particular, X-ray transitions obey the dipole selection rules given in Table 5.1.

If an inner shell electron is removed from an atom by any process including high-energy electron impact, a vacancy is created. This inner shell vacancy can then filled by an outer electron jumping down (see Fig. 8.1). Such transitions give rise to characteristic X-ray emission lines.

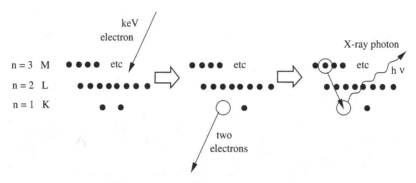

Fig. 8.1. X-ray line spectrum caused by the removal of an inner shell electron by electron impact followed by the subsequent relaxation of an outer shell electron with emission of an X-ray photon. This is how X-rays are usually produced in the laboratory.

Transitions involving K shell ($n = 1$ electrons, see Table 4.1) involve most energy: 'hard' X-rays or about 10^5 eV for heavy atoms. These energies correspond to wavelengths, λ, of about 0.1 Å. Higher shells give progressively 'softer' X-rays, with wavelengths up to tens of Å. Wavelengths between about 100 Å and 1000 Å are often described as the 'extreme ultraviolet' or EUV.

In fact, most X-rays produced in astronomical sources are not associated with particular transitions but are part of the continuum spectrum produced in high-energy environments such as active galactic nuclei (AGN). For example, it is thought that AGN generate X-rays from the atoms being accreted by a central massive black hole. These spectra can be extremely bright, producing more radiation than the rest of the entire galaxy; they give direct information on the distribution of accreted material. The solar corona provides a nearby source of X-ray spectra. Spectral lines come largely, not from inner shells of neutral atoms, but from highly ionised atoms. See Fig. 8.2 for example.

Consider the spectrum of Fe XXVI as an example. Fe^{25+} is a one-electron system with $Z = 26$. This means that its spectrum is similar to hydrogen but with all energies and transition frequencies scaled by Z^2. For H, the Lyα photon has an energy of 10.2 eV, while for Fe XXVI, Lyα has an energy of $10.2 \times (26)^2$ eV $\simeq$ 7 keV. Emission of Fe XXVI Lyα can be seen in the X-ray emission coming from intergalactic plasma which lies within clusters of galaxies. The continuum emission from this plasma is thermal emission with a temperature in the region of 10^8 K. At this

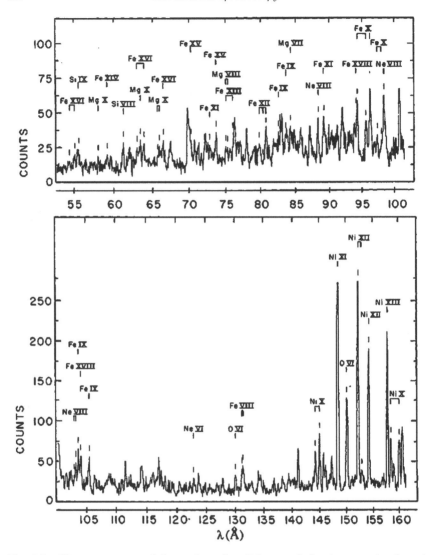

Fig. 8.2. X-ray spectrum of the entire solar disk recorded using a rocket-based experiment. [Adapted from M. Malinovsky and L. Heroux, *Astrophys. J.* **181**, 1009 (1973).]

temperature, iron is predominantly ionised to Fe^{25+} (see Fig. 3.10). This spectral region also contains lines from other highly-ionised species such as Fe XXV (see Fig. 8.3). This figure also shows the spectral signature of Fe I Kα transitions. These transitions arise from ionisation of a 1s electron

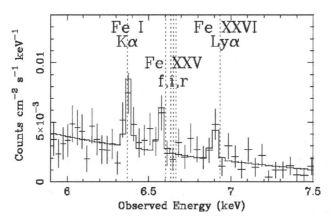

Fig. 8.3. X-ray spectrum of Seyfert galaxy NGC 7314 showing various iron transitions recorded using satellite Chandra [see T. Yaqoob *et al.*, *Astrophys. J.* **596**, 85 (2003)]. (T. Yaqoob, private communication.)

in Fe and then relaxation of a high-lying electron which emits at one of these characteristic wavelengths. As discussed above, such transitions have long been observed in the laboratory but had been thought to be unimportant astronomically until the recent batch of satellite-borne X-ray experiments.

Astronomical X-rays cannot be observed from ground. Early observations were made from rockets but now several satellites have been deployed. Recent space missions include SOHO and XMM-Newton launched by ESA and Chandra launched by NASA. Other space satellites are planned. These missions have substantially changed our view of the high-energy regions of the Universe as they have opened up many objects for detailed observations of X-ray spectra which previously could only be performed on the Sun, which has a very large flux of X-rays.

8.1 The Solar Corona

The visible spectrum of the solar corona shows a number of strong emission lines which for a long time could not be identified. These emissions were labelled 'coronium', although they were not generally thought to belong to a new element.

Seventy years after their original observation in 1869, the Swedish astronomer Bengt Edlen (1906–1993) showed that the lines were due to forbidden transitions in a number of highly-ionised atoms. The three most prominent lines are:

$$6374\,\text{Å}\ [\text{Fe x}]\ 3s^2 3p^5\ ^2P^o_{\frac{3}{2}} - {}^2P^o_{\frac{1}{2}}\,,$$

$$5302\,\text{Å}\ [\text{Fe XIV}]\ 3s^2 3p\ ^2P^o_{\frac{1}{2}} - {}^2P^o_{\frac{3}{2}}\,,$$

$$5694\,\text{Å}\ [\text{Ca XV}]\ 2s^2 2p^2\ ^3P_0 - {}^3P_1\,.$$

These transitions all occur within a single configuration and, indeed, within a single term. They are thus all strongly forbidden by electric dipole selection rules as they do not satisfy the Laporte rule. The transitions are all weak magnetic dipole fine structure lines. The green [Fe XIV] line at 5302 Å is the strongest of these coronal lines.

Transitions involving highly-ionised atoms such as Fe^{9+}, Fe^{13+} and Ca^{14+} will occur at high temperatures. Inspection of Fig. 3.10 suggests temperatures in the region of one million Kelvin. This temperature is much higher than the assumed temperature of the solar corona so Edlen's assignments were at first very controversial. However his analysis is now accepted but the mechanism for heating the corona remains not fully understood.

Many other coronal lines from highly-ionised metals can be observed in the solar spectrum in the extreme ultraviolet (EUV) and X-ray. Figure 8.4 shows the X-ray spectrum of the nearby star Capella, which is a G-star like the sun. It shows many emission features due to highly-ionised ions.

8.2   Isotope Effects

As discussed in Sec. 3.3, the energy levels of an atom depend on the reduced mass of the atom. For a hydrogen-like atom with nuclear mass M, the reduced mass is:

$$\mu = \frac{Mm_e}{M + m_e}\,. \tag{8.1}$$

Isotopes have the same atomic number Z and the same electronic structure, but have different nuclear masses, as expressed by the atomic weight, A. For example, deuterium is an isotope of hydrogen. In standard notation, the integer atomic mass (i.e. the total number of protons and neutrons) is given as a leading superscript on the atomic symbol. Thus hydrogen is ^{1}H and deuterium, often designated D, should more correctly be given as ^{2}H.

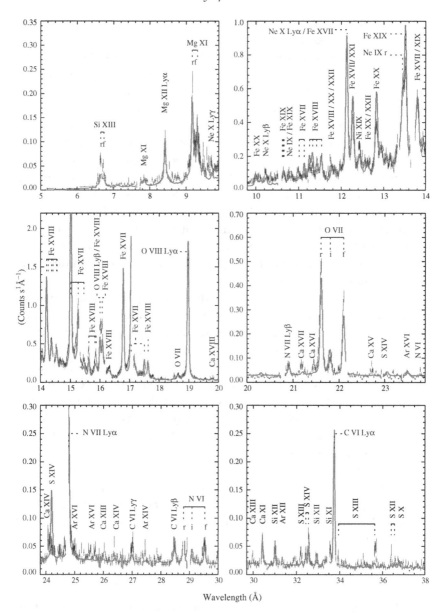

Fig. 8.4. X-ray spectrum of binary star Capella recorded using the XMM-Newton satellite [see M. Audard *et al.*, *Astron. Astrophys.* **365**, L329 (2001)]. (M. Audard, private communication.)

For H-like atoms, the energy levels and frequencies simply scale according to the reduced mass factor [see Eq. (3.8)]. For complex atoms the situation is not so simple. There is no longer a single scale factor as the mass effects of several electrons have to be allowed for simultaneously. The effect is also smaller than in hydrogen and immeasurably small for atoms with Z beyond about 30.

The measurement of the isotopic composition of a particular environment is important as it varies with the age of the Universe and the nuclear reactions occurring in a particular star. Isotopes thus provide an important measure of local nuclear reactions. For example, deuterium was formed in the early Universe but is burnt early in the life of all but lightest stars. The abundance of deuterium thus decreases with the age of the Universe and its presence can be used as an indicator that a brown dwarf lies below the critical mass necessary to burn D. Similar analyses can be conducted with other isotopes, although these are generally made rather than destroyed by stellar nuclear synthesis.

For most elements the best way of determining isotopic abundance is via molecular spectra; methods to do this are discussed in Sec. 10.1.1. For very heavy atoms, however, the effect of nuclear size becomes important.

So far we have always treated the nucleus as a point charge. Although very small, a factor of about 10^5 smaller than a typical atom, the nucleus does have a finite size. Furthermore the nucleus slightly distorts the electronic wavefunction and associated energy levels. This distortion is particularly important for electrons in s ($l = 0$) orbitals as these orbitals have nonzero densities at the nucleus (see Fig. 3.1) and for heavy elements which have larger nuclei.

An example is the element mercury, Hg, which has $Z = 80$, and so is heavier than iron and is therefore only present in significant quantities in chemically peculiar stars. Mercury can exist in one of seven stable isotopic forms with atomic weight A equal to 196, 198, 199, 200, 201, 202 and 204. Hg^+ has a prominent line:

$$3984 \text{ Å Hg II } 5d^9 6s^2 - 5d^{10} 6p \; ^2D_{\frac{5}{2}} - \; ^2P^o_{\frac{3}{2}} \, .$$

Note that this transition is formally forbidden as it is a two-electron transition, but in practice, strong configuration interaction (CI) effects ensure that it is actually strong. The change in the occupancy of the 6s orbital in this transition makes it sensitive to nuclear size effects, since all s orbitals have some probability of the electron overlapping the nucleus. When observed at very high resolution, the 3984 Å line can be separated into different wavelength components ranging from 3983.77 Å for ^{196}Hg$^+$ to 3984.07 for ^{204}Hg$^+$, allowing the isotopic composition to be determined. On earth the commonest isotopes of mercury are ^{200}Hg and ^{202}Hg; yet for no star are the terrestrial abundances of mercury isotopes observed. Instead, chemically peculiar stars show a range of isotopic compositions but with a tendency for the heaviest isotope ^{204}Hg to be the most abundant.

The analysis of mercury spectra discussed above is complicated by the fact that the spectrum of each isotope has its own hyperfine structure due to nuclear spin effects. Isotopes differ not only in their atomic weights but also in their nuclear spin, I. For example hydrogen, ^{1}H, is a single proton and therefore has $I = \frac{1}{2}$, but deuterium, ^{2}H, has $I = 1$. Similarly the nucleus of ^{4}He, which is an alpha particle, has zero spin, $I = 0$, but ^{3}He has $I = \frac{1}{2}$. For heavy elements, I can be larger and values up to 7 or 8 are found.

As discussed for hydrogen (see Sec. 3.14), coupling the nuclear spin, I, with the total (electronic) angular momentum, J, leads to the 'final' angular momentum quantum number, F:

$$\underline{F} = \underline{I} + \underline{J}. \tag{8.2}$$

Weak splittings of transitions according the value of F are called the hyperfine structure. The hyperfine structure gives isotopic information, but in practice, is only fully resolved in ultrahigh-resolution spectra and is therefore not generally astronomically useful. Use of such structure for different isotopes also requires very detailed laboratory data which is not always readily available.

Problems

8.1 Give an expression for the energy levels of the hydrogen-like atom of nuclear charge Z in terms of the Rydberg constant R_∞. Obtain the wavenumber of the Lyα transition of hydrogen-like oxygen, O^{7+}. From what astronomical environments would such transitions occur and how might they be observed?

8.2 Mercury has atomic number Z of 80. Give the full configuration of the $5d^96s^2$ ^{2}D$_{\frac{5}{2}}$ state of Hg$^+$. The ground state of Hg$^+$ has the configuration $5d^{10}6s$. Obtain the term and level for this state. Why is the $5d^96s^2$ ^{2}D$_{\frac{5}{2}}$ state metastable?

CHAPTER NINE

MOLECULAR STRUCTURE

'Ban dihydrogen monoxide'

– Spoof webpage *www.circus.com/~nodhmo/* (2004)

It is easy to record the astrophysical spectra of atoms and ions since they are present in the atmospheres of hot stars which shine brightly. Molecules, conversely, are found in cooler and less active regions which can make them more difficult to observe. Nonetheless, molecules can be found in many different astronomical environments. These range from planetary atmospheres and comets to cool stars and sunspots; from cold, giant interstellar clouds to planetary nebulae and even the tori which form round active galactic nuclei.

The spectra of molecules are studied at wavelengths from the ultra-violet to the radio. However it was the development of radio astronomy that led to the realisation that substantial portions of our galaxy, and hence the Universe, are dominated by molecular processes, indeed complicated ones. So far over 120 molecules have been detected in the interstellar medium by direct observation of their spectra. These molecules contain up to 13 atoms and their spectroscopy is rich and complex. There have also been claimed detections of significantly larger molecules, but these are harder to verify.

The introduction to molecular structure and spectroscopy presented here will largely be restricted to diatomic (two-atom) molecules as they show most of the key features of molecular spectra. For a discussion of the spectroscopy of larger (polyatomic) molecules, see Banwell and McGrath (1994), or Bernath (1995) in further reading. Both these textbooks give a more advanced and a more general treatment of the laboratory spectroscopy of molecules.

9.1 The Born–Oppenheimer Approximation

Molecules formed from two atoms are called *diatomics* while those formed from more than two atoms are described as *polyatomic*. Because molecular binding energies are relatively small, i.e. generally less than ionisation energies, molecules are only found in cooler, or less active, astronomical environments. Polyatomic molecules are only a significant component of matter at temperatures below about 4000 K. Diatomic systems can survive to somewhat higher temperatures and may be found in environments with temperatures up to about 8000 K. Of course, these figures are a rule of thumb with details depending on the molecule in question and the physical conditions of the environment.

The structure, and hence the spectra, of molecules are more complicated than atoms in two ways:

(1) There is no single charge centre about which electrons move. The electronic wavefunctions therefore have lower symmetry, making them harder to calculate and harder to work with.
(2) The nuclei themselves move, giving rise to both rotational and vibrational motions of the atoms within the molecule. These motions give rise to discrete spectra.

To a very good approximation, known as the *Born–Oppenheimer approximation*, the motions of the electrons and the nuclei can be completely separated. The validity of this approximation relies on the fact that the electrons are much lighter than the nuclei. For hydrogen, which is the worst case, $M_H \simeq 1836\,m_e$. This means that the electrons move very much faster than the nuclei and can be considered to relax instantly to any change in the positions of the nuclei. To use an everyday scale analogy, it is like flies buzzing round an elephant — as the elephant moves the flies move with it.

Within the Born–Oppenheimer approximation, one separates the wavefunction for the motions of electrons from the wavefunction for the motions of the nuclei [see Eq. (9.7)]. One can then consider the electronic wavefunction separately for each position of the nuclei, as if the nuclei are held fixed. The electronic energy associated with each of these electronic wavefunctions gives the familiar potential energy curves which show the interactions between two atoms. Such curves, examples of which are given in Fig. 9.2, only exist within the Born–Oppenheimer approximation.

To a less good approximation, one can also consider separately the two types of nuclear motion: vibration and rotation. The energies of these motions are such that the energy associated with motion of the electrons (the electronic energy) is always very much greater than the energy contained in vibrational motion which, in turn, is greater than the energy of rotation. This ordering is very useful when considering molecular structure but it is important to remember that in molecular spectra a particular transition can involve changes in more than one type of motion. For example, one gets rotational structure on vibrational transitions, and electronic transitions have vibrational structure and also fine structure due to simultaneous changes in rotational motion. The rotational fine structure in electronic transitions produces many lines close together. This fine structure can only be resolved at very high resolution which is often not available for astronomical spectra. If the rotational fine structure is not resolved, then the result is characteristic molecular band spectra.

9.2   Electronic Structure of Diatomics

Within the Born–Oppenheimer approximation, consider the electronic state of a diatomic molecule as a function of the separation, R, between the two atoms. Figure 9.1 shows typical coordinates for a diatomic, AB, where the atoms have nuclear charges Z_A and Z_B respectively. For clarity only two electrons are shown explicitly, so this figure can be taken to represent molecular hydrogen, H_2.

When considering the electronic state of the molecule AB as a function of R, the following considerations come into play:

As $R \to 0$, at very small R the overall interactions are strongly repulsive. There is repulsion due to nuclear–nuclear interaction

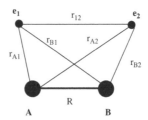

Fig. 9.1.   Coordinates for a diatomic molecule, AB, with two electrons.

whose potential depends on $\frac{Z_A Z_B}{R}$. There is also repulsion due to the electron–electron interactions, which also behave approximately as R^{-1}. However the most important interaction at small R in a many-electron molecule is due to exchange forces. Exchange interactions arise as a direct consequence of the Pauli Principle (see Sec. 4.3), which strongly forbids any attempt to force the two electrons to occupy the same space. This behaviour is similar to that found in collapsed stars where degeneracy pressure is a consequence of the Pauli Principle. It is the exchange forces that cause the atoms in molecules to behave almost as hard spheres when the molecule is compressed. The repulsion due to exchange forces at small R increases approximately as an exponential. It is therefore best modelled as e^{aR}, where a is a system-dependent constant.

As $R \to \infty$, the molecule is pulled apart and it separates into atoms in a process known as *dissociation*. The energy of the system at dissociation is clearly just the sum of the atomic energies.

At intermediate R, to get binding there must be some region of R where the molecular energy is less than the sum of the atomic energies. In this case the electronic state is described as 'attractive' and there is a minimum in the potential energy curve. The attraction can arise from interactions such as the increase in the electron–nuclear attraction brought about by bringing the atoms together. States can also be purely repulsive. Such states are referred to as 'dissociative' since they do not support bound states and will always lead to dissociation.

Figure 9.2 gives potential energy curves for the lower-lying electronic states of molecular hydrogen. The scheme used to derive labels for these curves is discussed in the next section. Note that the electronic potential is by convention denoted V. Two important quantities for the curves in Fig. 9.2 are the equilibrium bondlength and the dissociation energy. R_e, the equilibrium internuclear separation is defined by

$$\frac{dV}{dR}\bigg|_{R_e} = 0, \tag{9.1}$$

that is R_e is the value of R at the minimum of the potential. D_e is the dissociation energy of the molecule. It is the minimum energy required to fragment a molecule. For a diatomic molecule it can be defined by

$$D_e = V(R_e) - V(R = \infty). \tag{9.2}$$

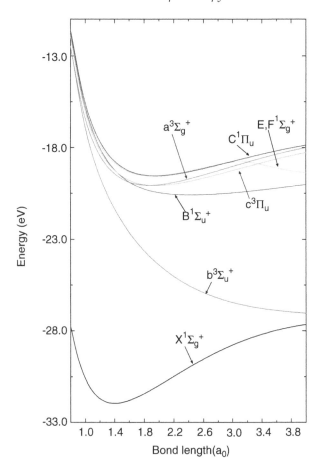

Fig. 9.2.   Potential energy curves for the lowest seven electronic states of molecular hydrogen, H_2. The lowest two curves both dissociate to two ground states, 1s hydrogen atoms; the higher curves dissociate to one H atom in its 1s ground state and the other into either a 2s or a 2p excited state. [Reproduced from D.T. Stibbe, PhD thesis, University of London (1997).]

9.2.1   *Labelling of electronic states*

Electronic states of atoms are labelled according to the so-called spectroscopic notation (see Sec. 4.8). The notation for electronic states of molecules follows a similar system but is more complicated because of their lower symmetry. In general terms the labelling scheme used is based on the symmetry of the molecule in question. Here only diatomic molecules will be considered. Of course, all diatomic molecules are linear molecules and only two separate symmetry cases need to be considered:

Homonuclear diatomics, where both atoms are the same, e.g. H_2, N_2, O_2.

Heteronuclear diatomics, where the atoms are different, e.g. CO, HF, CH.

As the electron spin is usually not strongly coupled to the frame of the molecule, the treatment of spin follows very much the same principles as in the atomic case. Each electron has its individual spin angular momentum, s_i. These can be summed to give a total spin angular momentum, S, again remembering that paired electrons in closed shells make zero contribution to this sum. The electronic states of molecules are designated by their spin multiplicity, $2S + 1$, which is given as a leading superscript, exactly as in the atomic case.

In an atom the treatment of the individual orbital angular momenta, l_i, follows along similar lines to the treatment of spin. However molecules are not spherical and the orbital angular momentum of the individual electrons is no longer a conserved quantity. For diatomic molecules, the total orbital angular momentum L is strongly coupled to the nuclear axis. It is therefore necessary to consider the components of L, designated Λ, along the diatomic nuclear axis which, by convention, is taken to define the z-axis of the system. What this means is that while the value of the total orbital angular momentum in a diatomic molecule can change, its projection onto the diatomic axis is conserved. As the projection of L onto z-axis can be either positive or negative, states with $\Lambda \neq 0$ are twofold degenerate while Σ states, which have $\Lambda = 0$, are singly degenerate.

Electronic states are labelled by their value of Λ rather than L. Values of Λ are denoted using the Greek letter equivalent of the Latin letter used to denote L (see Table 9.1), thus Σ, Π, Δ are equivalent to the atomic

Table 9.1. Letter designations for projected total orbital angular momentum quantum number, Λ.

$\Lambda =$	0	1	2	3	4 ...
Orbitals	σ	π	δ	ϕ	γ ...
States	Σ	Π	Δ	Φ	Γ ...
Degeneracy	1	2	2	2	2 ...

symbols S, P, D. Again, uppercase letters are used to denote many electron states and single electron orbitals are labelled using lowercase versions of the same symbols.

This means that:

$$^1\Sigma \text{ denotes a state with } S = 0 \text{ and } \Lambda = 0,$$
$$^3\Pi \text{ denotes a state with } S = 1 \text{ and } \Lambda = 1, \text{ etc.}$$

For most (stable) diatomics, the electronic ground state is a closed shell, meaning that it is $^1\Sigma$. Examples include H_2, N_2 and most other homonuclear diatomics. The exception is O_2 which has a $^3\Sigma$ ground state. CO and many other heteronuclear diatomics with an even number of electrons also have $^1\Sigma$ ground states. Diatomics with an odd number of electrons usually have $S = \frac{1}{2}$. For example H_2^+, CH^+ and CN all have $^2\Sigma$ ground states. Ground states need not have $\Lambda = 0$ (Σ), just as some atoms have ground states with $L > 0$. Examples include CH, OH and NO which all have $^2\Pi$ ground states. Molecules in electronic states with $\Lambda > 0$, and in particular those with $\Lambda = 1$, have extra lines in their spectra due to a process called Λ-doubling. This is a significant extra complication which will be discussed in Sec. 10.4 because of the astronomical importance of the CH and OH molecules. Indeed, CH was one of the first molecules to be detected outside our solar system.

9.2.2 *Symmetry*

There are two symmetry properties of certain diatomics which are similar to the parity of an atom. These contribute to electronic selection rules and hence are included in the electronic state designation.

Homonuclear diatomics have identical nuclei. This means that all properties must be unchanged with respect to swapping the two nuclei. The usual way of treating this is to consider the behaviour of the electronic wavefunction when the molecule is inverted through its centre-of-gravity.

Consider an electronic wavefunction, $\psi_e(\underline{R}_A, \underline{R}_B)$, where $\underline{R}_A$ is the vector connecting the molecular centre-of-gravity with nucleus A, and $\underline{R}_B$ connects the centre-of-gravity with nucleus B. As the two nuclei are identical the probability distribution of the electronic wavefunction must be unchanged with respect to inverting the molecule A – B → B – A, which has the effect of interchanging the nuclei:

$$|\psi_e(\underline{R}_A, \underline{R}_B)|^2 = |\psi_e(-\underline{R}_A, -\underline{R}_B)|^2. \qquad (9.3)$$

This equation has two solutions

$$\psi_e(\underline{R}_A, \underline{R}_B) = \pm\psi_e(-\underline{R}_A, -\underline{R}_B). \tag{9.4}$$

Both solutions are acceptable. Electronic states which are positive with respect to interchange are known as 'gerade', the German word for 'symmetric', or more simply 'g'. Electronic states whose wavefunctions change sign upon inversion are known as 'ungerade', or 'u'. The g/u label is given as a trailing subscript on the symbol used to designate the electronic state. Thus, for example, the ground state of H_2 is of $^1\Sigma_g^+$ symmetry while the next two stable states are $^1\Sigma_u^+$ and $^3\Pi_u$ which lie above the repulsive $^3\Sigma_u^+$ state. It should be noted that the g/u label applies not just to diatomic molecules but to all molecules with a centre-of-symmetry and that this label is not related to the total angular momentum of the system.

For Σ states only, there is an extra symmetry that needs to considered. This is the reflection of the wavefunction through any mirror plane containing the nuclei. Again, the electronic wavefunction Ψ_e either does or does not change sign with respect to this reflection. This parity is denoted by a trailing superscript for Σ states only, that is states are denoted either Σ^+ or Σ^-. Nearly all states are Σ^+, which are sometimes written simply as Σ. The one notable exception is the oxygen molecule which has a ground state configuration $1\sigma_g^2 1\sigma_u^2 2\sigma_g^2 2\sigma_u^2 3\sigma_g^2 1\pi_u^4 1\pi_g^2$ which gives the lowest electronic state of symmetry $^3\Sigma_g^-$. This reflection symmetry is hard to visualise since it requires two electrons in an open non-σ orbital, such as the $1\pi_u^2$ orbital in O_2, to get a Σ^- state.

The discussion above has focused on the overall label for the electronic state which is of course a many-electron property. Molecular orbitals, which are one-electron functions, have symmetry labels similar to those used for the states. Thus orbitals are labelled $\sigma, \pi, \delta, \ldots$ according to the projection, λ, of the (single-electron) orbital angular momentum onto the molecular axis. As in the atomic case, lowercase letters are used for one-electron properties. For homonuclear systems the orbital labels are subscripted g or u following the prescription given above. Thus the ground state configuration of H_2 is $1\sigma_g^2$. It should be noted that the σ-orbital does not have a $+$ or $-$ superscript. This is because a one-electron state can only have symmetric parity and all σ-orbitals are assumed to be $+$.

9.2.3 State labels

Giving an electronic state a symmetry designation such as $^1\Sigma_g^+$ is not unique. In fact there is likely to be an infinite number of states with this symmetry for a given molecule. In atoms it is common, if not universal practice, to precede the spectroscopic term value with n, the principal quantum number of the outer electron. For molecules this system would still not give a unique set of labels, so a different, rather more *ad hoc* system is used.

Each electronic state is proceeded by a letter. The following convention is used to assign the appropriate letter:

X labels the ground electronic state;
A, B, C, ... label states of same spin multiplicity as the ground state;
a, b, c, ... label states of different spin multiplicity to the ground state.

In principle, states are labelled alphabetically in ascending energy order. In practice there many exceptions. For example, the lowest triplet state of H_2 is the b $^3\Sigma_u^+$ with the a $^3\Sigma_g^+$ lying somewhat higher. Similarly there is a higher singlet state, which has a double minimum, labelled E, F $^1\Sigma_g^+$. The reasons for these various mislabellings are usually historical.

9.3   Schrödinger Equation

The Schrödinger equation for a diatomic molecule with nucleus A, of mass M_A and nuclear charge Z_A, and nucleus B, of mass M_B and nuclear charge Z_B, and N electrons can be written as

$$\left(-\frac{\hbar^2}{2M_A}\nabla_A^2 - \frac{\hbar^2}{2M_B}\nabla_B^2 - \frac{\hbar^2}{2m_e}\sum_{i=1}^N \nabla_i^2 + V_e - E \right) \Psi(\underline{R}_A, \underline{R}_B, \underline{r}_i) = 0, \quad (9.5)$$

where the first two terms are the kinetic energy operators for the motions of nuclei A and B respectively, the third term gives the kinetic energy operator for the electrons, and V_e is the potential. The potential is given by the various Coulomb interactions within the molecule:

$$V_e = \frac{e^2}{4\pi\epsilon_0}\left(-\sum_{i=1}^N \frac{Z_A}{r_{Ai}} - \sum_{i=1}^N \frac{Z_B}{r_{Bi}} + \sum_{i=2}^N\sum_{j=1}^{i-1} \frac{1}{r_{ij}} + \frac{Z_A Z_B}{R} \right). \quad (9.6)$$

These terms represent respectively the attraction of the electrons by nucleus A, the attraction of the electrons by nucleus B, the electron–electron repulsion and the nuclear–nuclear repulsion. The coordinates used in this expression are defined in Fig. 9.1.

Within the Born–Oppenheimer approximation, one writes the wavefunction as a product of electronic, ψ_e, and nuclear, ψ_n, wavefunctions:

$$\Psi(\underline{R}_A, \underline{R}_B, \underline{r}_i) = \psi_e(\underline{r}_i)\psi_n(\underline{R}_A, \underline{R}_B). \tag{9.7}$$

In this case the electronic wavefunction satisfies the simpler Schrödinger equation

$$\left(-\frac{\hbar^2}{2m_e} \sum_{i=1}^{N} \nabla_i^2 + V_e - E_e \right) \psi_e(\underline{r}_i) = 0. \tag{9.8}$$

This equation is solved separately for each value of the internuclear separation, R. The resulting eigenvalue, E_e, is the electronic energy at each R and gives the electronic potential $V(R)$ upon which the nuclei move.

The nuclear wavefunction satisfies the Schrödinger equation

$$\left(-\frac{\hbar^2}{2M_A} \nabla_A^2 - \frac{\hbar^2}{2M_B} \nabla_B^2 + V(R) - E \right) \psi_n(\underline{R}_A, \underline{R}_B) = 0, \tag{9.9}$$

where E, the eigenvalue, is the total energy of the system.

Solving the Schrödinger equation (9.9) to obtain the total energy E is actually not very useful. This is because the equation deals with all the motions of the nuclei: vibrations, rotations and the translation of the whole system through space. The energy of the translational motion gives a continuum and including it in the total energy would obscure the remaining discrete spectrum of energy levels. However Eq. (9.9) is a two-body equation which can be separated — just as in the H-atom problem — into an equation for the translational motion of the centre-of-mass of the system plus an equation for the internal motion of the molecule. For a diatomic, this latter equation represents the motion of one body in a 'central' potential, a potential which depends on the distance between the particles but not their orientation. The effective mass of this one-body problem is the reduced mass, μ, where

$$\mu = \frac{M_A M_B}{M_A + M_B}. \tag{9.10}$$

Unlike the H-atom problem (see Sec. 3.3), the reduced mass differs significantly from the mass of the individual nuclei. For example, for a homonuclear system, $M_A = M_B$ and $\mu = \frac{M_A}{2}$.

If one neglects the translational motion, the Schrödinger equation for nuclear motion becomes

$$\left[-\frac{\hbar^2}{2\mu} \nabla^2 + V(R) - E \right] \psi_n(\underline{R}) = 0, \tag{9.11}$$

where coordinate $\underline{R} = (R, \theta, \phi)$ is a vector giving the internuclear separation R plus the orientation (θ, ϕ) of the molecular axis relative to the laboratory z-axis.

The vibrational and rotational motion cannot be separated rigorously. However such a separation, which gives a good first approximation, can be written as

$$\psi_n(\underline{R}) = \psi_v(R)\psi_r(\theta, \phi). \tag{9.12}$$

Considering first the angular or rotational motion, one obtains

$$\left\{ -\frac{\hbar^2}{2\mu R^2} \left[\frac{1}{\sin\theta} \frac{d}{d\theta} \left(\sin\theta \frac{d}{d\theta} \right) + \frac{1}{\sin^2\theta} \frac{d^2}{d\phi^2} \right] - E_r \right\} \psi_r(\theta, \phi) = 0, \tag{9.13}$$

where the differential, kinetic energy operator is the angular part of the Laplacian operator ∇^2, which has been written using spherical polar coordinates.

Equation (9.13) is a Schrödinger equation with a zero-potential. It is actually the same equation as the angular part of the H-atom problem. The solutions are therefore the same ones satisfied by the angular wavefunctions of the H atom:

$$\Psi_r(\theta, \phi) = Y_{JM_J}(\theta, \phi), \tag{9.14}$$

where Y_{JM_J} is a spherical harmonic (see Fig. 3.2), and

$$E_r = \frac{\hbar^2}{2\mu R^2} J(J+1), \tag{9.15}$$

where the rotational angular momentum quantum number J takes integer values, $J = 0, 1, 2, \ldots$. The projection of J onto the molecular axis is denoted M_J. It takes values $M_J = -J, -J+1, \ldots 0, \ldots J-1, J$. There are a total of $2J + 1$ possible values of M_J, which represents the degeneracy of the rotational state. As in atoms, the M_J states split in the presence of a magnetic field.

If the molecule is taken to rotate as a rigid body, which is the simplest approximation to molecular rotation, then R can be fixed at some value such as R_0. Within this rigid rotor model, the moment of inertia of the molecule is $I_0 = \mu R_0^2$ and one can write the rotational energy expression as

$$E_r = \frac{\hbar^2}{2I_0} J(J+1) = B_0 J(J+1), \tag{9.16}$$

where B_0 is known as the rotational constant of the molecule. Note the subscript denotes a vibrational level [see Eq. (9.30)]. In practice, molecules are not rigid: atoms move apart as a molecule rotates faster (J increases), giving rise to centrifugal distortion terms which can be used to correct the expression (9.16).

It should be noted that if $J = 0$ then $E_r = 0$. This means that there is no rotational zero point energy.

Measurement of the rotational spectrum of a molecule gives a value for B_0. The rotational constant B_0 can in turn be used to give an accurate determination of the average bondlength of the molecule. Laboratory spectroscopy, particularly of pure rotational spectra, is the main source of accurate data on the geometric structure of molecules.

The radial equation arising from separating Eq. (9.11) is

$$\left[-\frac{\hbar^2}{2\mu} \frac{d^2}{dR^2} + V(R) - E_v \right] \psi_v(R) = 0, \tag{9.17}$$

where, for simplicity, the dependence on the rotational energy E_r has been neglected. Within this assumption the energy of the system is given by

$$E = E_v + E_r. \tag{9.18}$$

In Eq. (9.17), $V(R)$ is not a simple function of R, so the equation has no general algebraic solution. However, for low values of E_v, $V(R)$ can be approximated by a parabola

$$V_{H0}(R) = V_0 + \frac{1}{2} k(R - R_e)^2, \tag{9.19}$$

where the curvature of the potential about its minimum is given by

$$k = \frac{d^2 V}{dR^2} \bigg|_{R=R_e}, \tag{9.20}$$

where k is known as the force constant. This parabolic form of the potential is the one which gives simple harmonic motion. Figure 9.4 compares a harmonic potential with a more realistic one.

Setting the zero of energy at the minimum of the potential well, $V_0 = V(R = R_e) = 0$, Eq. (9.17) becomes

$$\left[-\frac{\hbar^2}{2\mu} \frac{d^2}{dR^2} + \frac{1}{2}k(R - R_e)^2 - E_v \right] \psi_v(R) = 0 . \tag{9.21}$$

This is the quantum mechanical equation of the harmonic oscillator which is solved in most introductory texts on quantum mechanics. See Rae (2002) in further reading for example.

The quantum mechanical solution for energy levels of the harmonic oscillator model is

$$E_v = \hbar\omega \left(v + \frac{1}{2} \right) , \tag{9.22}$$

where the vibrational quantum number v takes integer values $v = 0, 1, 2, \ldots$, meaning that the vibrational levels supported by a harmonic potential are evenly spaced (see Fig. 9.3). In Eq. (9.22) ω is the angular frequency. It is standard to quote $\hbar\omega$ in wavenumber units of cm^{-1}. The

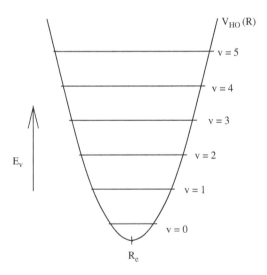

Fig. 9.3. Evenly-spaced energy levels supported by a harmonic potential. Note the zero point energy.

frequency is related to the constants in Eq. (9.21) by the expression

$$\omega = \left(\frac{k}{\mu}\right)^{\frac{1}{2}}. \tag{9.23}$$

The dependence on the reduced mass μ of this expression is important for studies of molecules with different isotopic composition. This will be discussed in Sec. 10.1.1.

It should be noticed that the vibrating molecule cannot exist at the bottom of the potential well. The minimum energy is called the *zero point energy* (zpe). For a diatomic harmonic oscillator the zero point energy equals $\frac{1}{2}\hbar\omega$. This zero point energy is a direct consequence of Heisenberg's Uncertainty Principle relating the position and the momentum. Because the potential has a minimum, a state with zero energy would be located precisely at R_e with no energy and hence no momentum; this would violate the uncertainty condition

$$\Delta x \Delta p_x \geq \frac{\hbar}{2}. \tag{9.24}$$

The zero point energy also means that the actual dissociation energy, D_0, is less than D_e defined by Eq. (9.2) above. The difference is given by

$$D_0 = D_e - \text{zpe}. \tag{9.25}$$

In other words the amount of energy taken to pull a diatomic molecule apart is reduced by the amount of the zero point energy.

9.4 Fractionation

The zero point energy of molecules has an important influence on the chemistry of the cold interstellar medium (ISM). Consider the simple deuterium substitution reaction:

$$H_2 + D \leftrightarrow HD + H. \tag{9.26}$$

At high temperatures, the ratio between $n(\text{HD})$ and $n(\text{H}_2)$ will be similar to that of $n(\text{D})$ to $n(\text{H})$, where $n(X)$ is the number density of species X. At the very low temperatures, below 20 K, found in giant molecular clouds in the ISM, this is not true due to zero point energy effects.

Consider the above example. The harmonic frequency of H_2 is $4396\,cm^{-1}$, implying a zpe of $2198\,cm^{-1}$. Mass scaling using Eq. (9.23) gives

$$\frac{\omega(HD)}{\omega(H_2)} = \left(\frac{\mu_{H_2}}{\mu_{HD}}\right)^{\frac{1}{2}}, \tag{9.27}$$

if one assumes that the force constant k does not change with isotopic substitution, which is true within the Born–Oppenheimer approximation. However

$$\mu_{H_2} = \frac{M_H}{2}, \quad \mu_{HD} = \frac{M_H M_D}{M_H + M_D} \simeq \frac{2}{3}M_H.$$

Assuming $M_D \simeq 2M_H$ and substituting into Eq. (9.27) gives

$$\omega_{HD} \simeq \omega_{H_2}\left(\frac{M_H}{2} \times \frac{3}{2M_H}\right)^{\frac{1}{2}} = \frac{\sqrt{3}}{2}\omega_{H_2} = 3807\,cm^{-1}.$$
$$zpe(HD) = \frac{1}{2} \times 3807 = 1903\,cm^{-1}. \tag{9.28}$$

A more accurate treatment gives $\omega_{HD} = 3817\,cm^{-1}$ and a zpe of $1909\,cm^{-1}$. The energy change for reaction (9.26) is given by

$$\Delta E = zpe(HD) - zpe(H_2) \simeq -289\,cm^{-1} \simeq 420\,K.$$

The lower zero point energy of HD compared to H_2 means that at low temperature, this reaction, and other similar ones, strongly favour HD formation over H_2. This effect, which can lead to between a hundred and 10000 times more D being observed in molecules than would be expected on abundance grounds, is termed 'deuterium fractionation'. It is observed in many species, including H_2/HD, HCO^+/DCO^+ and H_3^+/H_2D^+.

One can get fractionation of other isotopes such as ^{13}C over ^{12}C. However, the effect is less extreme as the mass differences are smaller.

9.5 Vibration–Rotation Energy Levels

Within the harmonic oscillator approximation all energy levels are evenly spaced. Real molecules are not harmonic oscillators. Their potential is elastic but the repulsion at short bondlengths is stronger than the attraction at long bondlengths, so a parabola is only the first approximation to $V(R)$.

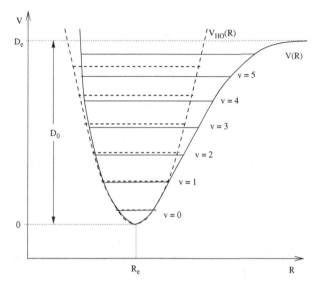

Fig. 9.4. Comparison of a harmonic potential (dashed line) with a more realistic one (full line) for a diatomic molecule with a fairly shallow potential well.

This approximation is reasonable near R_e but becomes increasingly poor at larger displacements which correspond to higher energies.

Figure 9.4 gives a qualitative comparison between the levels in a harmonic oscillator potential and those supported by a more realistic potential, one which actually leads to dissociation. Potentials which are not harmonic are referred to as *anharmonic*. Unlike the evenly-spaced harmonic energy levels, the energy levels get closer together as v increases. Also, the harmonic oscillator potential supports an infinite number of energy levels, whereas in reality the discrete levels stop at dissociation. Thus, for example, H_2 supports 14 vibrational levels in its ground electronic state. The precise number of levels depends on the well depth, curvature and the reduced mass involved.

A more complete expression for the energy of a vibration–rotation state above the bottom of the well for a $^1\Sigma$ ($S = 0, \Lambda = 0$) diatomic molecule is given by

$$E_{\mathrm{VR}} = \omega_e \left(v + \frac{1}{2} \right) + B_e J(J+1) - \omega_e x_e \left(v + \frac{1}{2} \right)^2$$
$$- D_e J^2(J+1)^2 - \alpha_e \left(v + \frac{1}{2} \right) J(J+1) + \cdots, \tag{9.29}$$

where the signs are chosen so that each constant, defined below, is positive. In Eq. (9.29), the expansion constants, known as spectroscopic constants, are:

ω_e,   the harmonic frequency;
B_e,   the rotational constant;
$\omega_e x_e$,   the leading anharmonic correction;
D_e,   the centrifugal distortion (not the dissociation energy);
α_e,   which gives variation of B_e with vibrational state.

Thus the rotational constant for vibrational state v is given by

$$B_v = B_e - \alpha_e \left(v + \frac{1}{2} \right) + \cdots . \tag{9.30}$$

The constants are all usually positive and given in cm^{-1}. They are usually determined empirically by analysing observed laboratory spectra and are different for each electronic state of a given molecule. Huber and Herzberg (1979) provide an extensive tabulation of these constants for the observed electronic states of diatomic molecules.

Values of spectroscopic constants for H_2, CO and the molecular ion CH^+ are given in Table 9.2. It is notable that even for hydrogen which is a relatively non-rigid system, the energy expressions, at least for low levels of excitation, are dominated by the harmonic oscillator, rigid rotor terms. In other words, the use of only ω_e and B_e gives quite a good representation of the actual energy levels. It should also again be noted that the rotational energies are much smaller than vibrational energies.

Table 9.2. Spectroscopic constants for the vibration–rotation levels of the ground electronic state of molecular hydrogen, carbon monoxide, and the CH^+ molecular ion. Values, which are all in cm^{-1}, are taken from Huber and Herzberg (1979).

Constant	H_2	CO	CH^+
ω_e	4401.21	2170	2740
$\omega_e x_e$	121.33	13.5	64.0
B_e	60.853	1.93	14.177
D_e	0.0471	6×10^{-6}	0.0014
α_e	3.062	0.017	0.492

9.6 Temperature Effects

Before considering temperature effects in detail it is useful to give a summary of the energy scales found in molecules. Figure 9.5 gives the energies in the H_2/H_2^+ system. Note that the rotational energy differences are too small to be visible on the scale given.

As molecules have many closely-spaced energy levels, they act as a useful thermometer in a variety of environments. To understand this it is necessary to consider the population of the various levels in a typical molecule as a function of temperature.

9.6.1 *Rotational state populations*

Consider the Boltzmann distribution for a diatomic molecule rotating as a rigid rotor. For simplicity take the $J = 0$ level as the zero of energy. Define

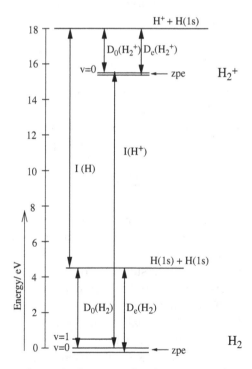

Fig. 9.5. Energy scales in hydrogen molecule. Zero is taken as the $H_2(v = 0)$ ground state. I represents the ionisation potential and D the dissociation energy.

$P(J)$ as the probability of finding the molecule in level J. To avoid having to evaluate the partition function, let us consider the ratio of $P(J)$ to the probability of finding the molecule in its rotational ground state, $P(0)$:

$$\frac{P(J)}{P(0)} = \frac{g_J}{g_0} \exp\left(\frac{-E_J}{kT}\right) , \tag{9.31}$$

where g_J is the degeneracy factor which is $(2J + 1)$ for a heteronuclear diatomic. Note that for homonuclear diatomics there is an additional factor caused by nuclear spin effects and ortho/para states, but this is beyond the scope of the present book. The energy levels of a rigid rotor are $E_J = BJ(J + 1)$, therefore

$$\frac{P(J)}{P(0)} = (2J + 1) \exp\left[\frac{-BJ(J + 1)}{kT}\right] . \tag{9.32}$$

For a typical diatomic, $B \simeq 1\,\mathrm{cm}^{-1}$, and in these units it useful to remember that Boltzmann's constant, k, equals $0.695\,\mathrm{cm}^{-1} \cdot \mathrm{K}^{-1}$. For a temperature of $T = 300\,\mathrm{K}$, these values give

$$\frac{P(1)}{P(0)} = 3 \exp\left(\frac{-1 \times 2}{0.695 \times 300}\right) = 3e^{-0.0096} = 2.97 ,$$

which means that there are more molecules in the $J = 1$ level than in the $J = 0$ level. In fact $P(0) \simeq P(28)$ for this example.

For a diatomic rigid rotor with rotational constant B, the rotational level with the largest population is given by the approximate relationship:

$$J = \left(\frac{kT}{2B}\right)^{\frac{1}{2}} - \frac{1}{2} . \tag{9.33}$$

At room temperature many rotational levels of a molecule are occupied. Even at cold interstellar medium temperatures of 10–20 K, most molecules have several levels occupied. Indeed, one of the earliest molecules observed in the interstellar medium, CN, was found in the early 1940s. The Nobel Prize-winning molecular spectroscopist Gerhard Herzberg (1904–1999) commented on this work that 'from the intensity ratio of the lines... a rotational temperature of 2.3 K follows, which has of course a very restricted meaning'. It is not as restricted as he thought, as Arno Penzias and Robert Wilson received a Nobel Prize for their 1963 demonstration of the existence of the cosmic microwave background radiation with a temperature of 2.7 K. The CN rotational temperature measurement was actually the first observational evidence of the cosmic microwave background radiation. The spectrum of CN is now used as one means of measuring this background temperature in different locations.

The rotational level populations of molecules such as CO and CN can be used to determine temperature in a particular region. However in many diffuse environments, the rotational levels of heteronuclear diatomics, or indeed any molecule with a permanent dipole moment, are usually not thermally occupied as the radiative lifetime, given by A_{if}^{-1}, is less than the mean time between collisions. That means the molecules exist in an environment which is below their critical density. The molecular emissions in planetary nebula NGC 7027 (Fig. 7.6) are like this.

9.6.2 *Vibrational state populations*

Vibrational energy separations are larger than rotational ones and are therefore only sensitive to higher temperatures. To illustrate this, consider the probability, $P(v)$, of finding a diatomic molecule in vibrational level v. For simplicity, assume that the diatomic is a harmonic oscillator and set the zero of energy at the $v = 0$ level. Again taking the ratio of the vth state to the vibrational ground state gives

$$\frac{P(v)}{P(0)} = \exp\left(\frac{-\hbar\omega v}{kT}\right). \tag{9.34}$$

It should be noted that there are no degeneracy factors associated with the vibrational motion.

Diatomic vibrational frequencies range from $500\,\text{cm}^{-1}$ for a heavy system to $4000\,\text{cm}^{-1}$ for H_2. $1000\,\text{cm}^{-1}$ is a typical value so taking this for a temperature of $300\,\text{K}$ gives

$$\frac{P(1)}{P(0)} = \exp\left(\frac{-1000}{0.695 \times 300}\right) \simeq 0.0083,$$

which means that less than 1% of the molecules are in their $v = 1$ vibrational level.

So at low temperatures, molecules are vibrationally cold with only the vibrational ground state significantly occupied. However, hot environments can lead to thermal emissions from vibrationally excited molecules.

This is quite often found for H_2 such as in Figs. 7.6 and 10.10 for example.

9.6.3   *Electronic state populations*

For most molecules the dissociation energy is less than the lowest electronic excitation energy. This means that electronically-excited states are not thermally occupied as the molecules are destroyed instead. Emission from electronically-excited states are still observed but these emissions are usually not thermal. For example, emissions are observed in planetary aurora, from electronically-excited H_2 in Jupiter and Saturn, and from other diatomics on Earth. These emissions are the direct consequence of excitation by collisions with electrons.

Astronomically there is one important exception to this rule. The atmospheres of cool stars, such as M-dwarfs, which have temperatures in the 2200–3700 K range, contain molecules. In these stars the most important diatomic molecules are H_2 and CO. However, a number of molecules, such as TiO, FeH, ZrO and VO, are formed which contain a transition metal atom. These molecules have open shells which result in many low-lying electronic states. These states are thermally occupied in the atmospheres of cool stars. These molecules have dense spectra which are often a significant source of opacity as they absorb radiation from the interior of the star over a wide range of wavelengths. The spectra of the molecules are very complicated. They are often poorly understood and in many cases remain completely uncharacterised. Figure 6.9 gives an example which shows that the absorptions in small, cool stars are dominated by molecular absorptions. The spectra of these open shell species will not be considered further here.

Problems

9.1 A sample of a diatomic molecule with rotational constant B, is in thermodynamic equilibrium at temperature T. Show that the ratio of the number of molecules in rotational level J to the number in rotational level zero is a maximum for the level with

$$J = \left(\frac{kT}{2B}\right)^{\frac{1}{2}} - \frac{1}{2},$$

where k is Boltzmann's constant, which has a value $0.695 \, \mathrm{cm}^{-1} \cdot \mathrm{K}^{-1}$.

9.2 For the molecule CO, $B_0 = 1.93\,\text{cm}^{-1}$. Which are the most-occupied rotational states of CO at

(a) the ISM temperature of $T = 20\,\text{K}$,
(b) room temperature, $T = 300\,\text{K}$,
(c) the temperature of a typical M-dwarf star of $3000\,\text{K}$?

Assuming a harmonic frequency $\omega(CO) = 2170\,\text{cm}^{-1}$, what proportion of the molecules are in the $v = 1$ vibrational state compared to $v = 0$ at each of these temperatures?

9.3 The fundamental vibrational frequency of the $^{12}CH^+$ molecular ion is $\omega = 2075.5\,\text{cm}^{-1}$. Estimate the zero point energy of $^{12}CH^+$ and, assuming integer values for the atomic masses, estimate the zero point energy of the less-abundant $^{13}CH^+$ ion. Why might one want to observe spectra of the less-abundant $^{13}CH^+$ species instead of $^{12}CH^+$? CH^+ ions are found in giant molecular clouds. How would you expect the ratio $n(^{13}CH^+) : n(^{12}CH^+)$ to compare to the ratio $n(^{13}C) : n(^{12}C)$, where $n(X)$ represents the number density of species X?

CHAPTER TEN

MOLECULAR SPECTRA

The astronomical spectra of molecules give rise to three distinct types of transitions, which have to be considered separately.

Pure rotational transitions lie at long wavelengths ranging from radio frequencies for heavy polyatomic molecules, through the microwave, to the far-infrared for light hydrogen containing diatomics.

Vibrational transitions are important at mid-infrared wavelengths.

Electronic transitions lie at similar wavelengths to the allowed transitions of neutral atoms: the visible and ultraviolet. The wavelengths quoted are only typical ones. In each case the precise spectral region depends on the spectroscopic constants of the molecule in question. Each of the transition types, and their associated selection rules are considered in turn below.

10.1 Selection Rules: Pure Rotational Transitions

For a diatomic to undergo a pure rotation transition, it must have a permanent dipole moment, μ. Actually for any molecule to have a dipole-allowed rotational spectrum it must have an asymmetric charge distribution giving a permanent dipole moment. Heteronuclear diatomics possess a permanent dipole moment but homonuclears, such as H_2, do not. Molecular hydrogen therefore does not have a dipole-allowed rotational spectrum.

The strength of a rotational transition depends on μ^2, therefore molecules with large permanent dipoles have intense transitions. This means that molecules such as sodium chloride, $Na^+ - Cl^-$, which have very large dipole moments as a result of the almost complete charge separations, have particularly intense rotational transitions. This has greatly aided the astronomical detection of such species even at very small concentrations or column densities (see Fig. 10.1).

For diatomic molecules in Σ symmetry electronic states, dipole-allowed transitions obey the following selection rule:

$$\Delta J = \pm 1. \tag{10.1}$$

Within the rigid rotor approximation it is therefore straightforward to obtain the energy of the rotational transitions:

$$\Delta E_{\text{rot}} = B_0 \left[J'(J' + 1) - J''(J'' + 1) \right], \tag{10.2}$$

where the standard convention for molecular spectra has been followed — that $'$ is used to denote the upper state and $''$ means the lower state.

As the selection rule gives $J' = J'' + 1$,

$$\Delta E_{\text{rot}} = 2 B_0 J'. \tag{10.3}$$

This means that the rigid rotor approximation leads directly to very regular pure rotational spectra where the transitions are regularly spaced by $2B$. Thus the transition $J' - J'' = 1 - 0$ is at $2B$, $2 - 1$ at $4B$, $3 - 2$ at $6B$, and so forth. It should be noted that, in contrast to atomic transitions, molecular transitions are generally denoted *upper – lower*. Sometimes an arrow is inserted so that $J' \leftarrow J''$ denotes absorption and $J' \rightarrow J''$ denotes emission.

Centrifugal distortion leads to a slight narrowing of the gap between neighbouring transitions as J increases, however the regular progression is still clearly seen. Figure 10.2 gives a sample pure rotational spectrum of CO recorded in the laboratory. The regular spacing of the spectrum for both the most-abundant species, $^{12}C^{16}O$, and isotopically-substituted species is easily seen. Astronomical pure rotational spectra are, of course, also regularly spaced when plotted using a frequency scale. However it unusual for a single astronomical spectrum in the radio or far-infrared to span a wide range. The more usual representation for series of astronomical rotational lines are given in Fig. 10.1 for NaCl and Fig. 10.3 for CO.

Carbon monoxide, CO, is a particularly important species for astronomical observations. CO is the most stable diatomic molecule. It has a dissociation energy D_0 of 11.1 eV, which is more than double the D_0 value found for most other diatomics. As a result, in astronomical environments where molecules form, C and O usually combine to form CO,

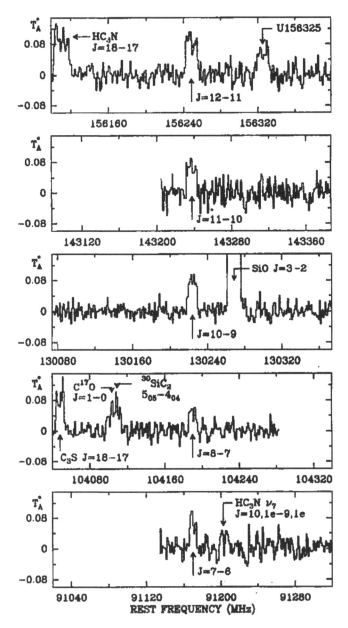

Fig. 10.1. Five spectra observed towards IRC+10216 with the IRAM 30 m telescope showing the $J = 7-6$ to $J = 12-11$ rotational transitions of NaCl. [Reproduced from J. Cernicharo and M. Guélin, *Astron. Astrophys.* **183**, L10 (1987).]

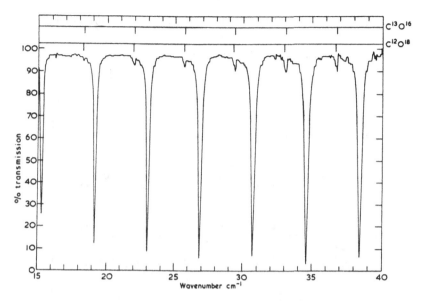

Fig. 10.2.   Laboratory far-infrared absorption spectrum of CO. [Reproduced from P.F. Bernath, *Spectra of Atoms and Molecules* (Oxford University Press 1995).]

which is very stable and long-lived. Any surplus C or O is then available for other chemistries. CO is thus the second most abundant molecule in the Universe after H_2.

For CO, the rotational constant of the ground vibrational state, B_0, is $1.93\,\mathrm{cm}^{-1}$. The wavelengths of the first few rotational transitions are $1-0$ at $\lambda = 2.60\,\mathrm{mm}$, $2-1$ at $1.30\,\mathrm{mm}$, and $3-2$ at $0.87\,\mathrm{mm}$. Examples of low J transitions are shown in Fig. 10.3. Much higher rotational levels can also be observed in warm objects such as the planetary nebula NGC 7027 (see Fig. 7.6).

The $J = 1-0$ transition of CO is the second most important spectral line in radio astronomy after the hydrogen 21 cm line. CO is widely distributed in the interstellar medium and maps of the CO $J = 1-0$ transition are a standard tool for investigating the ISM. One reason for this is that cold H_2 is very difficult to observe directly because its pure rotational transitions are not only very weak but lie in the near-infrared where ground-based observations are not possible. The abundance of CO is therefore often used to estimate the total amount of molecular gas present in a given environment. It is generally assumed that the number density of CO, $n(CO)$, is approximately $10^{-4}\,n(H_2)$.

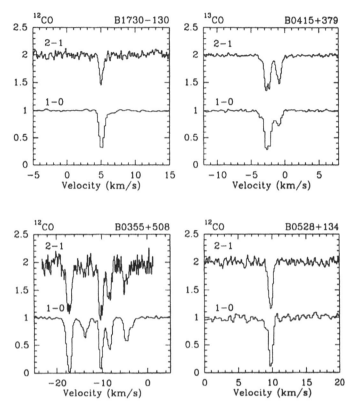

Fig. 10.3. Comparison of profiles from CO 2–1 and 1–0 absorption against extragalactic sources observed with the Plateau de Bure interferometer. The 2–1 spectra are displaced vertically by +1 for clarity. CO temperatures can be obtained by comparing the two absorption profiles. [Reproduced from R. Lucas and H.S. Liszt, in *Molecules in Astrophysics: Probes and Processes*, ed. E.F. van Dishoeck (Kluwer, 1997).]

If, as often happens, the CO 1–0 line is optically thick, one can use higher transitions such as the CO 2–1 line instead. These higher lines, at millimetre wavelengths, can be observed using telescopes such as the James Clerk Maxwell Telescope on Mauna Kea, Hawaii. Figure 10.3 gives an example of monitoring both the 1–0 and 2–1 lines together, which can be used for temperature determination. Another option to avoid the effects of optical thickness is to observe isotopically-substituted CO, which is present with much lower densities and whose transitions are therefore much less optically thick. As discussed below, isotopic substitution has a particularly marked effect on molecular spectra, making such observations fairly straightforward.

It should be noted that CO actually has a rather small permanent dipole of only 0.12 Debye. This means that even though the rotational spectrum is driven by this electric dipole, the Einstein A coefficient is actually fairly small for the pure rotational transitions of CO. For example $A_{1-0} = 7 \times 10^{-8}\,\text{s}^{-1}$. Note that the factor ν^{-3} in the definition of the Einstein A coefficient [see Eq. (2.4)], means that the radiative lifetime of rotationally-excited states is in any case significantly longer than those generally found for electronically-excited states of atoms or molecules whose transitions lie at much higher frequencies.

10.1.1 *Isotope effects*

Isotopic substitution means the replacement of an atom in a molecule by another isotope of that atom. Thus in CO, whose normal form is $^{12}\text{C}^{16}\text{O}$, one can replace ^{12}C by ^{13}C, giving $^{13}\text{C}^{16}\text{O}$. This change leaves the chemistry unaltered, except for the fractionation effects discussed in Sec. 9.4, but leads to changes in the spectrum due to mass effects.

The rotational constant for CO is

$$B_0 = \frac{\hbar^2}{2\mu r_0^2}, \quad \mu = \frac{M_C M_O}{M_C + M_O}. \tag{10.4}$$

Replacing ^{12}C by ^{13}C leads to an increase in the reduced mass μ and hence an increase in the moment of inertia, I, assuming that the effective bondlength r_0 is unchanged upon substitution. This leads to decreased rotational constant, B, for $^{13}\text{C}^{16}\text{O}$ and hence a smaller separation between the rotational transitions. This change in transition frequency is significant and can easily be resolved even with moderate resolution. Thus the $1-0$ transition is at 2.60 mm for $^{12}\text{C}^{16}\text{O}$, at 2.67 mm for $^{12}\text{C}^{17}\text{O}$, and at 2.72 mm for $^{13}\text{C}^{16}\text{O}$.

The spectrum of $^{13}\text{C}^{16}\text{O}$ can easily be observed. See Fig. 10.3 for example. In fact, astronomical spectra of all versions of isotopically-substituted CO have been recorded.

10.1.2 *Rotational spectra of other molecules*

There are presently over 120 different molecules whose spectra have been observed in the interstellar medium; this number is growing steadily by about 3 species per year. Most of these molecules have been detected by the observation of rotational transitions at radio frequencies. Radio spectra of giant molecular clouds are rich in features, particularly if one probes deeply by using long integration times and sensitive detectors. Figures 10.4 and 10.5 show a survey and a close-up spectrum recorded looking at the nearby massive star-forming region in Orion. Such studies

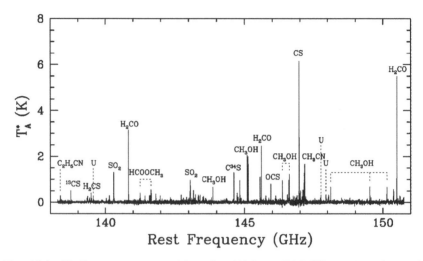

Fig. 10.4. Radio spectrum spanning the 138.3 to 150.7 GHz region obtained towards Orion-KL. Strong lines are designated their molecular identifications. 'U' marks unidentified lines. [Reproduced from C.W. Lee, S.-H. Cho and S.-M. Lee, *Astrophys. J.* **551**, 333 (2001).]

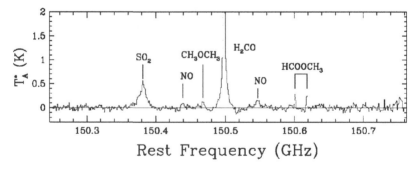

Fig. 10.5. Detailed radio spectrum spanning obtained towards Orion-KL. Lines are designated by their molecular identifications. [Reproduced from C.W. Lee, S.-H. Cho and S.-M. Lee, *Astrophys. J.*, **551**, 333 (2001).]

highlight the large difference in intensity scales of the various transitions that can be observed. That there are molecules still waiting to be found is beyond doubt because there are many unidentified spectral features. See for example the strong lines marked 'U' in Fig. 10.4. Higher-resolution studies reveal many, many more such features. Most often such features are assigned as a result of detective work based on spectroscopic studies performed in the laboratory.

Interstellar medium molecules are most often observed in emission, but can also be sometimes found in absorption. Indeed, there are a number of cases where molecules have been observed in absorption against the cosmic microwave background; these observations imply that the absorbing molecule is at a temperature below 2.7 K as measured by its rotational population. As in the atomic case, observation of high-resolution line profiles can yield further information: Doppler shifts can give relative motion, Doppler broadening can give the (translation) temperature, and so forth. Figure 10.6 gives an example of self-absorption by the carbon monosuphide molecule, CS. Light from further away is absorbed by the molecules nearer to us. In this case the nearer molecules are cooler as their absorption profiles are significantly narrower. These particular observations are interpreted as evidence for the collapse of the dark molecular cloud Barnard 335 as a precursor to star formation.

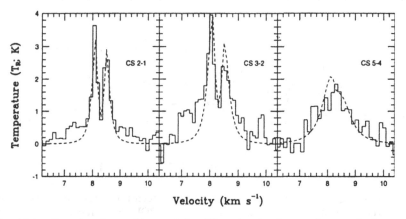

Fig. 10.6. Rotational transitions of the CS molecule observed towards the centre of massive dark cloud Barnard 335. The dashed lines are from a model fit. The $J = 2-1$ and $J = 3-2$ show self-absorbed profiles with stronger blue peaks. This behaviour is typical of in-falling gas. [Reproduced from S. Zhou, *Molecules in Astrophysics: Probes and Processes*, ed. E.F. van Dishoeck (Kluwer, 1997).]

10.1.3   *Rotational spectra of molecular hydrogen*

H_2 is a homonuclear diatomic and therefore does not have a permanent electric dipole moment or an allowed rotational spectrum. It should be noted that the molecule HD does have a dipole but only a very small one of 0.0008 Debye. This leads to an Einstein A coefficient of only $3 \times 10^{-8}\,\mathrm{s}^{-1}$ for the lowest $J = 1 - 0$ rotational transition.

However the abundance of H_2 in many locations in the Universe means that electric quadrupole transitions can be observed. These transitions are weak, like the forbidden quadrupole transitions found in atomic spectra. For diatomic molecules, the selection rule for pure rotational spectra driven electric quadrupoles are

$$\Delta J = \pm 2 . \tag{10.5}$$

The H_2 molecule is very light so that $B_0 = 62\,\mathrm{cm}^{-1}$, meaning that the rotational transitions of H_2 lie at infrared wavelengths. For example, the lowest $2 - 0$ line is at $\lambda = 28\,\mu$m. These transitions are about a billion times weaker than dipole-allowed rotational transitions. The Einstein A value for the $2 - 0$ transition is very small, $3 \times 10^{-11}\,\mathrm{s}^{-1}$, which corresponds to a radiative lifetime against spontaneous emission of about 1000 years.

H_2 quadrupole rotational transitions can be seen in the warm ISM in emission (see Fig. 10.7). Such transitions have been extensively studied in

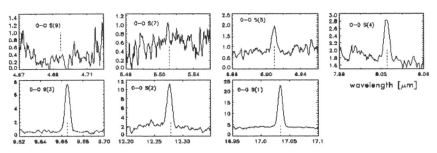

Fig. 10.7.   Quadrupole rotational emission spectra of molecular hydrogen towards the warm photon-dominated region (PDR) S140 recorded using the Infrared Space Observatory (ISO). The observations probe the warm, dense molecular gas adjacent to the S140 H II region. The flux density (vertical scale) is given in Jansky ($= 10^{-23}\,\mathrm{erg\,s}^{-1} \cdot \mathrm{cm}^{-2} \cdot \mathrm{Hz}^{-1}$). Dashed lines give the laboratory wavelengths for the transition indicated. [Adapted from R. Timmermann *et al.*, *Astron. Astrophys.* **315**, L281 (1996).]

Orion; indeed higher rotational transitions have been observed in Orion, up to $S(17)$ or $J = 19 - 17$, than those detected in the laboratory. In the laboratory, getting a sufficient pathlength of light through a sample of H_2 to observe these very weak transitions is a major challenge.

10.2 Vibrational Transitions

Within the harmonic oscillator approximation, there is a simple and rigorous electric dipole selection rule:

$$\Delta v = \pm 1. \tag{10.6}$$

This leads directly to

$$\Delta E_v = \hbar \omega \left(v + 1 + \frac{1}{2} \right) - \hbar \omega \left(v + \frac{1}{2} \right) = \hbar \omega, \tag{10.7}$$

where ω is known as the fundamental frequency.

For anharmonic molecules any change in Δv is allowed in principle, but in practice, $\Delta v = \pm 1$ always leads to much stronger transitions. The intensity of individual transitions falls off rapidly with increasing Δv. Vibrational transitions which change v by more than one quantum are generally called *overtones*.

Vibrational transitions are usually accompanied by a change in rotational state. For electric dipole transitions within a Σ electronic state, the additional selection rule on rotational motion is

$$\Delta J = \pm 1. \tag{10.8}$$

For non-Σ states and polyatomic molecules, transitions with $\Delta J = 0$ can also occur. To get dipole-allowed vibrational excitation, one must have a change in the dipole moment

$$\frac{d\mu}{dQ} \neq 0, \tag{10.9}$$

where Q is the vibrational coordinate undergoing excitation.

For homonuclear diatomics, $\mu = 0$ for all internuclear separations, so vibrational transitions are all dipole-forbidden. This is not true, however, for symmetric polyatomic linear molecules such as carbon dioxide.

CO_2 does not have a dipole moment in its equilibrium geometry but distortions from this can give an instantaneous dipole moment. Hence some CO_2 vibrational transitions are dipole-allowed. It is these vibrational transitions which contribute to the greenhouse effect both here on earth and on Venus, where the super-abundance of CO_2 in the atmosphere is responsible for maintaining a huge greenhouse effect.

For heteronuclear diatomics, such as CO, the derivative of the dipole with respect to bondlength, $\frac{d\mu}{dR}$, is only zero by accident. So heteronuclear diatomic molecules all have allowed vibrational spectra.

10.2.1 Structure of the spectrum

For simplicity the discussion will be confined to rotation–vibration spectra in Σ symmetry electronic states. Consider the so-called fundamental transition given by $v' = 1 - v'' = 0$ centred on the the fundamental frequency, ω. The rotational selection rules are $\Delta J = \pm 1$, meaning that the vibrational band has two branches. Assuming the rigid rotor approximation this gives:

R-branch: $\Delta J = +1$, i.e. $J' = J'' + 1$

$$\Delta E_{VR} = \hbar\omega + J'(J'+1)B_1 - J''(J''+1)B_0 \simeq \hbar\omega + 2B(J''+1) \quad (10.10)$$

for $J'' = 0, 1, 2, \ldots$. It has been assumed that $B_1 \simeq B_0$, which is generally a good approximation, and is about as reliable as neglecting the effects of centrifugal distortion.

P-branch: $\Delta J = -1$, i.e. $J' = J'' - 1$

$$\Delta E_{VR} = \hbar\omega + J'(J'+1)B_1 - J''(J''+1)B_0 \simeq \hbar\omega - 2BJ'' \quad (10.11)$$

for $J'' = 1, 2, 3, \ldots$ but not $J'' = 0$ as it would imply $J' = -1$. In standard notation the transitions for the two branches are labelled R(J'') and P(J''), where J'' is the rotational quantum number associated with the lower vibrational state.

This series of transitions give a characteristic spectral pattern as depicted in Fig. 10.8. The rotational selection rules lead to a series of lines

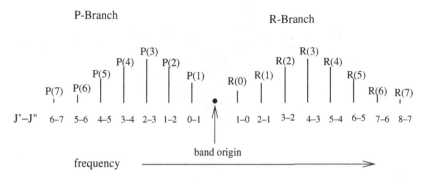

Fig. 10.8. Stick spectrum showing the structure of a vibrational band for a diatomic molecule with a Σ symmetry electronic state.

which are evenly spaced, $2B$ apart. The exception is at the centre of the band where no line is observed.

Figure 10.9 shows a CO absorption spectrum recorded looking towards the massive young stellar object AFGL 4176, which is embedded in a dense molecular cloud. The symmetric structure of the P- and R-branches, and the band centre, which can be pinpointed by the missing line, are all clearly visible. The shorter wavelength feature due to solid carbon dioxide shows how different these largely structureless condensed phase spectra are. The 4.7 μm region is heavily obscured by the earth's atmosphere. This spectrum was recorded by a satellite, the Infrared Space Observatory (ISO).

Besides the structure of the lines, the vibration–rotation spectra of diatomics also have a characteristic shape. The intensity of individual lines is directly proportional to the rotational population in the initial state of the molecule. This means that this intensity distribution can be used to measure the (rotational) temperature of the molecule. The temperature can then be estimated from a spectrum using formula (9.33), although a more reliable result can usually be obtained by fitting the whole spectrum to a temperature-dependent rotational population.

For diatomic molecules not in a Σ electronic state, that is, ones with $\Lambda > 0$, and for polyatomic molecules, the rotational selection rule (10.8) is generalised to

$$\Delta J = 0, \pm 1, \quad \text{not } J = 0 \leftrightarrow 0. \tag{10.12}$$

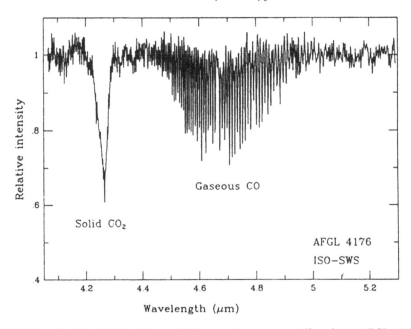

Fig. 10.9.   Spectra obtained towards the massive young stellar object AFGL 4176 using the Infrared Space Observatory (ISO). The strong, largely structureless band absorption at 4.27 μm is due to solid CO_2, whereas the characteristic vibration–rotation P- and R-branch structure between 4.4 and 4.9 μm is due to the presence of warm, gaseous CO along the line of sight. [Reproduced from E.F. van Dishoeck, in *The Molecular Astrophysics of Stars and Galaxies*, eds. T.W. Hartquist and D.A. Williams (Clarendon Press, Oxford, 1998).]

The extra $\Delta J = 0$ transitions form what is known as a Q-branch. Within the rigid rotor approximation, all Q-branch transitions in a vibration–rotation spectrum lie at the same frequency, giving the spectrum a characteristic sharp peak at the band origin. Effects beyond the rigid rotor model can introduce slight shifts in these transitions but it is common for the individual Q-branch transitions to not be resolved even with high resolution.

Vibrational spectra can be seen in emission from hotter regions of the ISM and of planetary atmospheres particularly from auroral regions. They are seen in absorption in the atmospheres of cool stars and brown dwarfs. Absorption spectra of cold ISM molecules can be seen against the light of a suitable, more distant infrared source. Such observations, which are often necessary to identify interstellar molecules which do not have a permanent dipole, can be difficult.

10.2.2 *Isotope effects*

Isotopic substitution alters vibrational frequencies as well as rotational ones. However, whereas the rotational transition frequency shifts according to the inverse of the reduced mass μ (see Sec. 10.1.1), the vibrational band origin shifts as $\mu^{-\frac{1}{2}}$. This can be seen from Eq. (9.23) which is based on the harmonic approximation. A further assumption, valid within the Born–Oppenheimer approximation, is that the force constant, k, does not change with isotopic substitution. Thus for $^{13}C^{16}O$ the band origin is shifted to $2096\,\mathrm{cm}^{-1}$ from $2143\,\mathrm{cm}^{-1}$ found for $^{12}C^{16}O$. Such shifts are sufficient to be detectable with a spectrograph of moderate resolution.

10.2.3 *Hydrogen molecule vibrational spectra*

Like its rotational spectrum, H_2 only undergoes very weak electric quadrupole-induced transitions for its vibrational spectrum. For these transitions, vibrational state changes with $\Delta v = \pm 1$ are still dominant, although transitions with $\Delta v > 1$ are observed astronomically. The rotational selection rules for these quadrupole transitions are:

$$\Delta J = 0, \pm 2, \quad \text{not } J = 0 \leftrightarrow 0. \tag{10.13}$$

This leads to a new set of labels for the different branches. Transitions with $\Delta J = 2$ form the S-branch; $\Delta J = 0$ form the Q-branch; $\Delta J = -2$ form the O-branch. Transitions are labelled $S(J'')$, $Q(J'')$ and $O(J'')$, where J'' is the rotational quantum number of the lower rotational state involved in the transition.

Despite the extreme weakness of the transitions — they are approximately 10^9 times weaker than electric dipole-allowed vibration–rotation transitions — emissions from vibrationally excited H_2 are bright from a number of astronomical sources. These include planetary atmospheres (Fig. 10.10), planetary and reflection nebulae (Fig. 10.11) and hot regions of the ISM. H_2 emissions can be particularly bright from merging galaxies, which contain huge quantities of hot hydrogen molecules.

Both Figs. 10.10 and 10.11 show effects due to the earth's atmosphere, the so-called telluric effects. This is a common problem with observing in the infrared. In particular, the gap between 1.8 and 2.0 μm in Fig. 10.11 due to atmospheric effects separates the H-band (1.4–1.8 μm) from the K-band (2.0–2.4 μm).

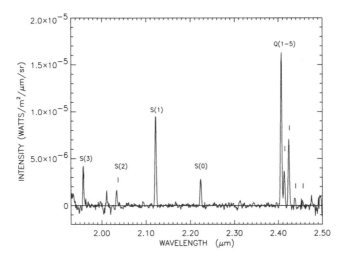

Fig. 10.10. Continuum-subtracted, infrared K-band spectrum of Uranus recorded using the UK Infrared Telescope (UKIRT) by Trafton and co-workers [see *Astrophys. J.* **405**, 761 (1993)]. The H_2 quadrupole emission lines are prominent, especially in the Q-branch. Weak emission features from the overtone band of the hydrogenic molecular ion H_3^+ can also be seen. The structure in the spectrum near 2.01 μm and long-wards of 2.45 μm is caused by telluric features.

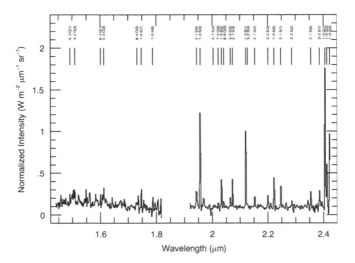

Fig. 10.11. Infrared spectra of reflection nebula NGC 7023. The wavelengths of molecular hydrogen lines have been labelled above the spectrum. The intensity has been normalised to the peak 1 – 0 S(1) intensity. [Reproduced from P. Martini, K. Sellgren and J.L. Hora, *Astrophys. J.* **484**, 296 (1997).]

10.3 Electronic Transitions

Electronic transitions involve a jump from one potential energy curve to another curve (see Fig. 10.12). The most common situation is that the equilibrium bondlength of the excited electronic state, R'_e, is bigger than that of the ground state, R''_e, as electronic excitation usually weakens the bonding in a molecule. The potential energy curves of H_2 (see Fig. 9.2), illustrate this behaviour.

Electronic spectra are significantly more complicated than the pure rotational and vibration–rotation spectra considered above. This is because it is necessary to consider changes in the electronic, vibrational and rotational state of the molecule for each possible change in electronic state.

10.3.1 *Selection rules*

Nearly all electronically-excited states of molecules contain enough energy to dissociate the molecule. For this reason it is usually only necessary to consider electronic transitions which obey electric dipole selection rules. Table 10.1 gives the selection rules for electric dipole transitions

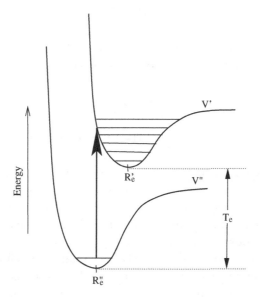

Fig. 10.12.   Electronic transitions involve a jump from one potential to another.

Table 10.1. Selection rules for spectra of diatomic molecules undergoing allowed electric dipole transitions.

Rotations	$\Delta J = \pm 1$	for $\Lambda = 0 - 0$,
	$\Delta J = 0, \pm 1$	not $J = 0 - 0$, for other $\Delta \Lambda$.
Vibrations	Δv any.	
Spin	$\Delta S = 0$.	
Orbital	$\Delta \Lambda = 0, \pm 1$.	
Σ states	$\Sigma^+ \leftrightarrow \Sigma^+$,	
	$\Sigma^- \leftrightarrow \Sigma^-$.	
Symmetry	$g \leftrightarrow u$	Homonuclear molecules only.

for diatomic molecules. In principle these selection rules apply to all transitions of a diatomic molecule, but in practice the table is of most use for considering electronic transitions as these have the most complicated selection rules. The simplified versions of the selection rules given above for pure rotation and vibration–rotation transitions are sufficient for these cases.

For all molecules, allowed electric dipole transitions obey the spin conserving selection rule

$$\Delta S = 0. \tag{10.14}$$

For diatomic molecules the projection of the total electron orbital angular momentum along the diatomic axis, Λ, obeys the rule:

$$\Delta \Lambda = 0, \pm 1, \tag{10.15}$$

which means that $\Sigma - \Sigma$, $\Sigma - \Pi$ and $\Pi - \Delta$ transitions are allowed but a $\Sigma - \Delta$ transition is forbidden.

Molecules with a centre-of-symmetry, which include homonuclear diatomics, must change symmetry when undergoing an electric dipole transition. Thus transitions must link g and u electronic states; transitions which are $u \leftrightarrow u$ and $g \leftrightarrow g$ are forbidden. This rule is the molecular equivalent of the Laporte rule encountered in atomic spectra. Its application explains why there are no dipole-allowed rotation or vibration–rotation spectra for homonuclear diatomics since such transitions all occur within a single electronic state and therefore leave the symmetry of the state unchanged.

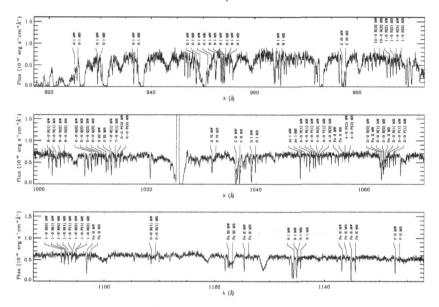

Fig. 10.13.   Spectrum towards the star DI 1388 recorded with the Far-Ultraviolet Space Explorer (FUSE). The interstellar atomic and molecular lines arising in the Galaxy (MW) and the Magellanic Bridge (MB), a region between the large and small Magellanic clouds, are identified. [Reproduced from N. Lehner, *Astrophys. J.* **578**, 126, (2002).]

Figures 10.13 and 10.14 show ultraviolet spectra of molecular hydrogen recorded by observing starlight passing through the interstellar medium. Prominent electronic transitions shown for H_2 are the

Werner band: $C\,^1\Pi_u - X\,^1\Sigma_g^+$ at about 1100 Å;
Lyman band: $B\,^1\Sigma_u^+ - X\,^1\Sigma_g^+$ at about 1010 Å.

These transitions are described as bands and do not occur at a precise wavelength because within each electronic transition there are a series of vibrational and rotational transitions. These are considered in turn below.

10.3.2   *Vibrational selection rules*

There are no rigorous selection rules which govern the change in vibrational quantum numbers during an electronic transition. The chance of a particular vibrational transition occurring depends on the squared overlap

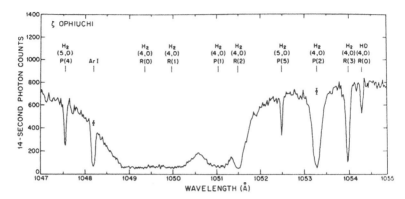

Fig. 10.14. Spectrum recorded towards O-star ζ Oph using the satellite Copernicus. [Reproduced from L. Spitzer Jr. and E.B. Jenkins, *Ann. Rev. Astron. Astrophys.* **13**, 133 (1975).]

of the vibrational wavefunctions from the two potentials involved:

$$\left| \int \psi'_v(v')^* \psi''_v(v'') dR \right|^2 . \tag{10.16}$$

This factor, which takes a value between zero and one, is called a Franck–Condon factor. It is derived following the Franck–Condon principle, which can be thought of as an extension of the Born–Oppenheimer approximation as it assumes that electronic transitions occur too rapidly for the nuclei to move during them. See Bransden and Joachain (2003) in further reading for a full discussion of the Franck–Condon principle.

It is common practice to denote vibrational bands within an electronic transition by (v', v''). So $(4, 0)$ means a transition between vibrational state $v' = 4$ in the upper state and $v'' = 0$ in the lower state.

10.3.3 *Rotational selection rules*

The rotational selection rule for an electronic transition is

$$\Delta J = 0, \pm 1, \quad \text{not } J = 0 \leftrightarrow 0, \tag{10.17}$$

except that $\Delta J = 0$ transitions are not allowed for transitions between Σ electronic states. These selection rules lead to P-branch $(J'-J'' = -1)$, Q-branch $(J' = J'')$ if allowed, and R-branch $(J' - J'' = +1)$ transitions. However, unlike vibrational spectra, these branches do not display a regular structure.

Table 10.2. Spectroscopic constants, in cm^{-1}, for the electronic states of molecular hydrogen involved in the Lyman and Werner bands.

State	T_e	ω_e	$\omega_e x_e$	B_e	α_e	D_e
$C\,^1\Pi_u$	100089.8	2443.77	69.524	31.362	1.644	0.0223
$B\,^1\Sigma_u^+$	91700.0	1358.09	20.888	20.015	1.185	0.0163
$X\,^1\Sigma_g^+$	0	4401.21	121.33	60.853	3.062	0.0471

The rotational constant associated with each electronic state involved in a transition is different, i.e. $B' \neq B''$; in fact they are often substantially different. This is because although the reduced mass of the system stays constant during a transition, the equilibrium bondlength, R_e, does not. The most common situation is that $R_e' > R_e''$ so that $B' < B''$. Table 10.2 gives the molecular constants which can be used to determine the wavelength of individual transitions in the Werner and Lyman bands. This shows that B_e for the $X\,^1\Sigma_g^+$ ground state is substantially bigger than B_e for the two electronically-excited states.

The difference in rotational constants means that the rotational 'fine structure' lines are not evenly spaced. Furthermore the transitions often pile up on top of each other leading to a band head where many rotational transitions occur very close together. Band heads occur at the red end of the band, if $B' < B''$, which is the most common situation, or at the blue end of the band, if $B' > B''$.

10.3.4 *Transition frequencies*

The energy difference, ΔE, for an electronic transition contains contributions from each of the electronic, vibrational and rotational energies. The difference is most conveniently written

$$\Delta E = T_e + E_{VR}' - E_{VR}'' , \qquad (10.18)$$

where T_e is the adiabatic electronic excitation which is defined as the energy difference between the minima in the potential energy curves of the two states involved (see Fig. 10.12). By convention the minimum of the ground state curve is taken as the energy zero. The vibration–rotation energy levels within each curve, E_{VR}, are usually obtained in terms of standard molecular constants. Expressions for them are given in Eq. (9.29). The relevant constants for the Lyman and Werner bands of H_2 are given in Table 10.2; a general tabulation of spectroscopic constants for diatomic molecules can be found in Huber and Herzberg (1979).

Worked Example: A cold interstellar column of H_2 absorbs light in the B $^1\Sigma_u^+ - X\,^1\Sigma_g^+$ Lyman band. This band does not have a Q-branch. Assuming that the H_2 molecule is entirely in its vibrational ground state and $J'' = 0$ and 1 rotational states, what wavenumber of light will be absorbed to populate vibrational states with $v' = 0$ and 1 in the upper electronic state?

Solution: Denote the energy levels $E_X(v'', J'')$ and $E_B(v', J')$ for the ground and excited electronic states respectively. Then, using spectroscopic constants for the appropriate electronic state taken from Table 10.2,

$$E_X(0,0) = \frac{w_e''}{2} - \frac{w_e'' x_e''}{4} = 2170.3\ \text{cm}^{-1},$$
$$E_X(0,1) = E_X(0,0) + 2B_e'' - 4D_e'' - \alpha_e'' = 2288.8\ \text{cm}^{-1},$$
$$E_B(0,0) = T_e + \frac{w_e'}{2} - \frac{w_e' x_e'}{4} = 92690.1\ \text{cm}^{-1},$$
$$E_B(0,1) = E_B(0,0) + 2B_e' - 4D_e' - 2\alpha_e' = 92726.5\ \text{cm}^{-1},$$
$$E_B(0,2) = E_B(0,0) + 6B_e' - 36D_e' - 6\alpha_e' = 92798.9\ \text{cm}^{-1},$$
$$E_B(1,0) = E_B(0,0) + w_e' - 2w_e' x_e' = 93721.5\ \text{cm}^{-1},$$
$$E_B(1,1) = E_B(0,1) + w_e' - 2w_e' x_e' - \alpha_e' = 93762.0\ \text{cm}^{-1},$$
$$E_B(1,2) = E_B(0,2) + w_e' - 2w_e' x_e' - 3\alpha_e' = 93842.7\ \text{cm}^{-1}.$$

The conditions specified only give six transitions. They are:

$$(0,0)\ \text{R}(0) \text{ at } \omega = E_B(0,1) - E_X(0,0) = 90242.3\ \text{cm}^{-1},$$
$$(0,0)\ \text{P}(1) \text{ at } \omega = E_B(0,0) - E_X(0,1) = 90085.0\ \text{cm}^{-1},$$
$$(0,0)\ \text{R}(1) \text{ at } \omega = E_B(0,2) - E_X(0,1) = 90200.9\ \text{cm}^{-1},$$
$$(1,0)\ \text{R}(0) \text{ at } \omega = E_B(1,1) - E_X(0,0) = 91156.2\ \text{cm}^{-1},$$
$$(1,0)\ \text{P}(1) \text{ at } \omega = E_B(1,0) - E_X(0,1) = 91401.3\ \text{cm}^{-1},$$
$$(1,0)\ \text{R}(1) \text{ at } \omega = E_B(1,2) - E_X(0,1) = 91510.1\ \text{cm}^{-1}.$$

10.3.5 *Astronomical spectra*

Most astronomical electronic spectra of molecules are observed in absorption. Recent observations by the Far-Ultraviolet Space Explorer (FUSE) satellite have extensively monitored the Werner and Lyman bands of H_2 in absorption in the ISM. Such observations provide a direct observational handle on cold molecular hydrogen in the ISM. These bands are observed in absorption against starlight. See Fig. 10.13 for example.

Figure 10.14 shows part of the Lyman band of H_2. This transition is $B\,^1\Sigma_u^+ - X\,^1\Sigma_g^+$ and is electric dipole-allowed. Absorption for the vibrational ground state starts at about 1100 Å and runs to shorter wavelengths. The abundance of H_2 is so great in interstellar clouds that absorption from the lowest rotational level, $J = 0$, is optically thick. Despite the large B constant of H_2, and the relatively small Einstein A coefficient for the transition, absorption from several rotationally excited levels can be seen. The spectrum also shows a line due to HD, which is shifted due to mass effects that influence the vibration–rotation energies in both electronic states. Deuterium fractionation effects at cold temperatures mean that although a ratio of HD to H_2 can be obtained from this spectrum, this does not give directly the ratio of D to H abundances.

Electronic spectra can be observed in emission from both comets and planetary aurorae. In both these cases the emissions are the result of non-thermal excitations. For example, Jupiter's aurorae emit strongly in both the Lyman and Werner bands as a result of electrons travelling down the magnetic field lines in the polar regions and exciting the H_2 gas which makes up most of Jupiter's atmosphere.

Except for H_2, whose rotational fine structure transitions are widely spaced, the rotational structure of the electronic transitions are often not fully resolved, leading to band structures. However the bands have characteristic profiles, for example containing band heads, which allow them to be identified as molecular in origin.

Molecules in giant molecular clouds and other cold interstellar environments do not possess sufficient energy for their electronic spectra to be observed in emission. However, molecular absorptions can be seen against suitable light from a more distant star. As these molecules exist at low temperatures, only the ground vibrational state and a few rotational levels are occupied. For light molecules only one rotational state may be occupied. These species give rise to fairly simple spectra with transitions to several vibrational states of the upper electronic state, each with a few, maybe even only one, rotational fine structure line.

The first three molecules observed in the ISM were CH, CH^+ and CN. These somewhat unusual species were all detected by their electronic spectra which lie in the visible. Figure 10.15 compares term diagrams, which can be thought of as the molecular equivalent of Grotrian diagrams, for these three species. CH and CH^+ are light species so only transitions from their rotational ground states are usually observed. The CN 'violet'

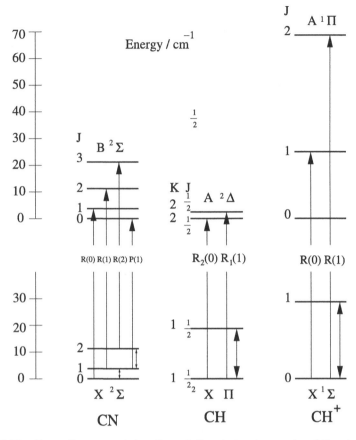

Fig. 10.15. Term diagrams giving the rotational structure only of the strongest electronic interstellar bands of the CN, CH and CH^+ molecules. Note the more complicated structure of CH is caused by Λ-doubling in both the ground and excited electronic states. [Reproduced from P. Thaddeus, *Ann. Rev. Astron. Astrophys.* **10**, 305 (1972).]

band has a larger transition dipole and CN is heavier and so has smaller rotational constants. Two or three rotational levels may be occupied, which leads to either three or five rotational fine structure transitions for each vibrational band. Figure 10.16 gives an example where only the $J'' = 0$ and 1 levels of CN are occupied. The population of CN rotational levels are sensitive to the local temperature of the environment and can be used to measure the local microwave background, as discussed in Sec. 9.6.1.

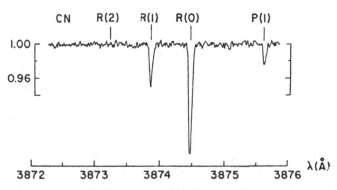

Fig. 10.16. Optical spectrum recorded towards ζ Oph showing rotationally-resolved absorption due to the CN 'violet' band. Absorption by the R(2) line is not observed. Adapted from P. Thaddeus [*Ann. Rev. Astron. Astrophys.* **10**, 305 (1972).]

10.4 Non-$^1\Sigma$ Electronic States

Molecules which do not have $^1\Sigma$ (i.e. $S = 0$, $\Lambda = 0$) electronic states have more complicated spectra. In this case it is necessary to consider the coupling of the rotational motion to the spin and/or orbital angular momenta.

This can get fairly complicated as, in particular, there is more than one way to do this. The choice of coupling scheme depends on the molecular species and even the electronic state in question. These different coupling schemes are known as *Hund's cases*. Here only one astronomically, and atmospherically important species will be considered: the hydroxyl molecule, OH.

The ground state of OH is X $^2\Pi$ which means that $S = \frac{1}{2}$ and $\Lambda = 1$. This system is treated using Hund's case (a) which starts by coupling the spin projection along the molecular axis, Σ, to the electron orbital projection on this axis. The resulting projection of the total electron angular momentum, Ω, can take two values as this is not vector addition but the addition of two projections which either point in the same direction or opposite directions. The total projection can thus take the values

$$\Omega = |\Lambda - \Sigma|, \Lambda + \Sigma. \tag{10.19}$$

For the case of $^2\Pi$ electronic state, therefore, Ω equals $\frac{1}{2}$ and $\frac{3}{2}$.

The X $^2\Pi$ electronic state of OH is split into two states which are designated $^2\Pi_{\frac{3}{2}}$ and $^2\Pi_{\frac{1}{2}}$, or indeed sometimes $^2\Pi_+$ and $^2\Pi_-$. The separation between these two states is due to relativistic interactions and therefore completely equivalent to the fine structure found in atomic levels. For OH the $^2\Pi_{\frac{3}{2}}$ is the lower of the two states.

The extra complication for molecular spectra arises from the additional coupling to the rotational motion which is strictly represented by the quantum number N. Note that N equals J in the usual case when $S = \Lambda = 0$. However for states with $\Lambda \neq 0$, the electronic angular momentum couples with N to give J. This coupling to the rotational motion splits each of the energy levels into a '+' or '−' component. The result is known as Λ-*doubling*. Λ-doubling effects are particularly strong for molecules such as OH which have $\Lambda = 1$. The structure of the low-lying energy levels of OH, and the resulting transitions between them, is given in Fig. 10.17.

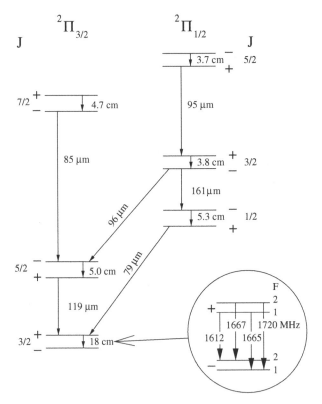

Fig. 10.17.   Low-lying energy levels of the OH molecule showing the wavelengths of allowed radio and near-infrared transition. The expanded portion shows the hyperfine splitting of the lowest levels and the frequencies of the associated hyperfine transitions.

Transitions between the Λ-doublets are allowed. The lowest of these lies at radio frequencies with a wavelength of 18 cm. Observation of this transition in 1963 led to OH becoming the fourth molecule observed in the interstellar medium and the second species observed at radio frequencies.

To add one more layer of angular momentum coupling, the Λ-doublets are themselves split by hyperfine coupling, which couples the rotational motion quantum number J to the total nuclear spin quantum number I. In the case of OH, ^{16}O has a zero nuclear spin and the hyperfine effects arise from coupling J to the $i = \frac{1}{2}$ of the H nucleus. Hyperfine effects cause the 18-cm line to split into four components (see the inset in Fig. 10.17). These lines can be observed in high-resolution studies.

10.5 Maser Emissions

Microwave amplification stimulated emission of radiation (maser) action is observed from at least 36 molecules including SiO, OH and water, usually at infrared or microwave frequencies. The population inversion necessary to cause maser action can be created by a number of mechanisms including optical pumping, radiation trapping in certain long-lived levels and selective collisional excitation of the masing molecule.

Figure 10.18 depicts a simplified case of a maser driven by collisional excitation. In this figure the levels are vibration–rotation levels of the electronic ground state. If level A is excited in some collision process, such as

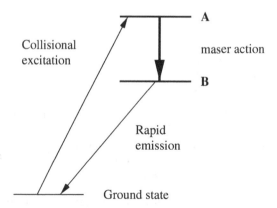

Fig. 10.18. Typical scheme for a three-level masing system driven by selective collisional excitation.

scattering with H_2, and level B is not, then the population of level A can be greater than the population of level B. This situation gives a non-thermal population and can lead to maser action which, by its nature, is highly directional.

The detailed mechanism which causes masing in each system can be regarded as an accident of the physics for that species. Molecules such as SiO have been observed masing in over 400 stars. OH and water are observed to mase in the ISM. Such maser action is often very time variable and — since masing is very dependent on the detailed physics — carries a lot of information about the local environment. Indeed the original observations of what turned out to be OH masers were so puzzling that they were dubbed 'mysterium'. Instead of being a possible new element, it was considered possible that these emission lines with strong, wildly fluctuating intensities were due to an alien lifeform. However it was demonstrated that simple physical phenomena giving maser action could give rise to such effects. The masing hyperfine lines of OH are now widely studied. Figure 10.19 gives an example of maser emission from a variable late-type supergiant star. Water and SiO maser lines can both be observed from this object. The velocity structure of the masing line displayed in Fig. 10.19 indicates that the emissions are coming from an expanding shell of gas at constant velocity.

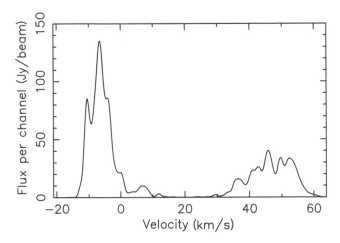

Fig. 10.19. Velocity profile of OH 1612 MHz maser emission from the variable, late-type supergiant star VX Sagitarii, observed using the Lovell 76 m antenna. (A.M.S. Richards, private communication.)

Problems

10.1 Interstellar CO is observed absorbing infrared radiation at 2134.63, 2138.46, 2146.16, 2150.03 and 2153.94 cm^{-1}. Identify the transitions observed in this spectrum and explain what astrophysical information might be obtained from its study. Where might one expect to observe an infrared spectrum of CO in (a) absorption and (b) emission?

10.2 One wishes to observe the molecules ^{12}C^{16}O and ^{13}C^{16}O. How do (a) the $J = 1-0$ rotational transition frequency and (b) the $v = 1-0$ fundamental vibrational frequency depend on the reduced mass of the system? For ^{12}C^{16}O, the $J = 1-0$ transition lies at 3.86 cm^{-1} and the fundamental frequency lies at 2170 cm^{-1}. Assuming integer values for the atomic masses, estimate these quantities for ^{13}C^{16}O. Why might it be necessary to observe different transitions for ^{12}C^{16}O than ^{13}C^{16}O to determine their relative abundances?

10.3 Observations are planned of CO pure rotational emission spectra using the $J = 5-4$ and $J = 20-19$ transitions. Assuming CO is a rigid rotor with rotational constant $B = 1.93$ cm^{-1}, estimate the transition frequency of the two transitions. Explain which of these estimates will be more accurate.

10.4 The ^{12}CH$^+$ molecular ion has a rotational constant $B = 11.94$ cm^{-1} and a fundamental vibrational frequency of $\omega = 2075.5$ cm^{-1}. A cold, interstellar column of ^{12}CH$^+$ in its $J = 1$ rotational state is observed in (a) the far-infrared and (b) the mid-infrared. Give the quantum numbers and wavenumber (in cm^{-1}) of the transitions you would expect to observe.

10.5 Assign quantum numbers transitions labelled CO in Fig. 7.6. You can assume these are all pure rotational transitions.

10.6 Figure 10.9 shows a vibrational spectrum of CO. Estimate the wavelength of the band origin of CO from this spectrum. Starting at the band origin, assign rotational quantum numbers as far as the maxima in the two branches of the spectra. Hence obtain an estimate of the temperature of the CO gas involved in the absorption. You may assume that CO is a rigid rotor with a rotational constant $B = 1.93$ cm^{-1}.

10.7 Estimate the wavenumber in cm^{-1} at which cold interstellar H$_2$ will absorb in the Werner band. Assume H$_2$ is a rigid rotor and a harmonic oscillator, and use the constants B_e, ω_e and T_e given in Table 10.2. Only consider absorption to the three lowest vibrational states, $v' = 0, 1, 2$, in the excited, C $^1\Pi_u$, electronic state.

SOLUTIONS TO MODEL PROBLEMS

Several solutions below give precise numerical answers to problems for which an estimate was requested. For these problems, indicated * below, correct working may yield answers differing slightly from those given.

Chapter 1

1.1 $v = c\frac{\Delta\lambda}{\lambda}$.

Line 1: $v = \frac{411.54 - 410.17}{410.17} \times 2.998 \times 10^8 = 100.1\,\mathrm{km\,s^{-1}}$;

Line 2: $v = \frac{435.50 - 434.05}{434.05} \times 2.998 \times 10^8 = 100.1\,\mathrm{km\,s^{-1}}$;

Line 3: $v = \frac{487.75 - 486.13}{487.75} \times 2.998 \times 10^8 = 99.9\,\mathrm{km\,s^{-1}}$;

Line 4: $v = \frac{658.47 - 656.28}{656.28} \times 2.998 \times 10^8 = 100.0\,\mathrm{km\,s^{-1}}$;

All lines show a Doppler shift of $100\,\mathrm{km\,s^{-1}}$ which is the speed the star is moving **away** from earth as the shift is positive.

Chapter 2

2.1* Doppler shift, measuring v_r as approximately $-600\,\mathrm{km\,s^{-1}}$, is

$$\Delta\nu = \frac{v_r}{c}\nu = \frac{6 \times 10^5}{3 \times 10^8} \times 99.023 = 0.198\,\mathrm{GHz}.$$

Hence $\nu = 99.023 + 0.198 = 99.221\,\mathrm{GHz}$. The laboratory frequency for H50β is $99.223\,\mathrm{GHz}$.

Chapter 3

3.1 $E_n = -\frac{R_H}{n^2}$. Lyα is $n = 2 - 1$ so

$$E_1 - E_2 = R_H\left(1 - \frac{1}{4}\right) = 109677.58 \times \frac{3}{4} = 82258.19\,\mathrm{cm^{-1}}.$$

R_∞ is defined for a one-electron atom with an infinite nuclear mass. Iron is a heavy atom, so the reduced mass for an electron plus an iron nucleus is very close to the mass of an electron, m_e, which is the mass used in defining R_∞. Lyα for Fe^{25+} therefore lies close to $26^2 \times 0.75 \times 109737.31 = 5.564 \times 10^7$ cm^{-1}. Such transitions will only be observed from very hot environments such as intergalactic plasma.

3.2 Selection rules: $\Delta l = \pm 1$, Δn any. Observed emissions:

$$4-3 \text{ P}\alpha \text{ at } 18751 \text{ Å};$$
$$4-2 \text{ H}\beta \text{ at } 4861 \text{ Å};$$
$$4-1 \text{ Ly}\gamma \text{ at } 972 \text{ Å};$$
$$3-2 \text{ H}\alpha \text{ at } 6563 \text{ Å};$$
$$2-1 \text{ Ly}\alpha \text{ at } 1215 \text{ Å}.$$

Starting from the 4s level, $3-1$ Lyβ at 1026 Å will be observed, but not Lyγ.

The cascade of emission from the H 4p level can be depicted as:

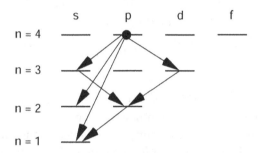

3.3 $\lambda^{-1} = \frac{R}{hc}\left(\frac{1}{n_1^2} - \frac{1}{n_2^2}\right)$ for Hα $n_1 = 2$, $n_2 = 3$. Only the Rydberg constant R changes between H and D, via the reduced mass, so:

$$\frac{\lambda_D}{\lambda_H} = \frac{R_H}{R_D} = \frac{\mu_H}{\mu_D} = \frac{M_H m_e}{M_H + m_e} \times \frac{M_D + m_e}{M_D m_e},$$
$$\frac{\lambda_D}{\lambda_H} = \frac{1836.1}{1837.1} \times \frac{3671.4}{3670.4} = 0.999728.$$

Therefore $\lambda_D(\text{H}\beta) = 6561.00$ Å.

$$R = \frac{\lambda}{\Delta\lambda} = \frac{\lambda_H}{\lambda_H - \lambda_D} = \frac{R_H^{-1}}{R_H^{-1} - R_H^{-1}} = \frac{1}{1 - R_H/R_D},$$

so the resolving power R is independent of n_1 or n_2.

$$R = \frac{1}{1 - 0.999728} \simeq 3700 \,.$$

Note that telescope resolving powers are usually given to one digit and a power of ten. It is meaningless to specify the resolving power to significantly more digits than this.

Since R is independent of n_1 and n_2, the same resolving power is required for Lyα. The main problem with observing both H and D transitions together is that if there is approximately 10^5 more H than D, then if the D transition is observable, the H one is likely to be optically thick.

3.4* From Sec. 3.8.1, the volume of an H atom with principal quantum number n is

$$V_n \approx \frac{4}{3} \pi n^6 \times 1.48 \times 10^{-31} \, \text{m}^3 \,.$$

(a) Star with $N = 10^{16} \, \text{cm}^{-3} = 10^{22} \, \text{m}^{-3}$, so each atom can occupy up to about $10^{-22} \, \text{m}^3$. Equating gives:

$$\frac{4}{3} \pi n^6 \times 1.48 \times 10^{-31} \simeq 10^{-22}$$

which gives $n \simeq 23$.

(b) H II region $N = 10^4 \, \text{cm}^{-3} = 10^{10} \, \text{m}^{-3}$, so each atom can occupy up to about $10^{-10} \, \text{m}^3$. Equating with V_n gives $n \simeq 2300$.

The partition function sum is finite since in any environment the H atom is confined in a finite volume. This truncates the sum at some maximum value of n, e.g. 23 for the star and 2300 for H II region considered above.

3.5* $\lambda^{-1} = R\left(\frac{1}{n_1^2} - \frac{1}{n_2^2}\right)$. For 80$\alpha$, $n_1 = 80$ and $n_2 = 81$ which gives $\lambda = 2.37668$ using R_∞ and $\lambda = 2.37797$ using R_H. The resolving power

$$R = \frac{\lambda}{\Delta\lambda} = \frac{2.377}{2.37797 - 2.37668} \approx 1800 \,.$$

A one-digit estimate of the resolving power gives $R = 2000$.

Chapter 4

4.1 Be has 4 electrons; its ground state configuration is $1s^2 2s^2$. Both orbitals are fully occupied, giving a term ^1S and a level $^1S_0$.

The excited state has configuration $1s^2 2s^1 3d^1$ giving $l(2s) = 0$, $s(2s) = \frac{1}{2}$ and $l(3d) = 2$, $s(3d) = \frac{1}{2}$. Adding these gives $L = 2$ and $S = 0$ or 1. These correspond to terms ^1D, ^3D, and levels $^1D_2$, $^3D_1$, $^3D_2$, $^3D_3$.

4.2 $4d^2$ means $l_1 = 2$, $s_1 = \frac{1}{2}$ and $l_2 = 2$, $s_2 = \frac{1}{2}$. Adding these $L = l_1 + l_2 = 0, 1, 2, 3, 4$ and $S = s_1 + s_2 = 0, 1$. Without the Pauli Principle the terms ^1S, ^1P, ^1D, ^1F, ^1G, ^3S, ^3P, ^3D, ^3F and ^3G would be possible, but the Pauli Principle imposes the rule that for indistinguishable electrons the sum of $L + S$ must be even. Applying this rule gives ^1S, ^3P, ^1D, ^3F and ^1G only. If Hund's rules are obeyed for this configuration then the energy order would be ^3F, ^3P, ^1G, ^1D, ^1S, going from lowest to highest.

Term	L	S	$J = \lvert L - S \rvert, \dots, L + S$
^1S	0	0	0
^3P	1	1	0, 1, 2
^1D	2	0	2
^3F	3	1	2, 3, 4
^1G	4	0	4

giving $^3F_2$, $^3F_3$, $^3F_4$, $^3P_0$, $^3P_1$, $^3P_2$, $^1G_4$, $^1D_2$, $^1S_0$. $^3F_2$ should be lowest in energy. All terms are even parity since $(-1)^{l_1+l_2} = (-1)^{2+2} = 1$.

4.3 $^4F^{\circ}_{\frac{7}{2}}$ means $S = \frac{3}{2}$, $L = 3$, $J = \frac{7}{2}$. It will have $2J + 1 = 8$ states (which are M_J sublevels). The other levels are given by $J = L + S = \frac{3}{2}, \frac{5}{2}, \frac{7}{2}, \frac{9}{2}$ which gives $^4F^{\circ}_{\frac{3}{2}}$, $^4F^{\circ}_{\frac{5}{2}}$ and $^4F^{\circ}_{\frac{9}{2}}$ in addition to $^4F^{\circ}_{\frac{7}{2}}$.

As $S = \frac{3}{2}$, the system must contain at least 3 electrons. A possible three-electron configuration could be 1s2p3d. Other configurations are also possible, but not $2p^3$ which cannot give $^4F^{\circ}$ by the Pauli Principle.

4.4 ^1D has $L = 2$ and $S = 0$ so it can exist only for $J = 2$ i.e. $^1D_2$, not $^1D_1$; $^0D_{\frac{5}{2}}$ would mean $2S + 1 = 0$, i.e. $S = -\frac{1}{2}$ but S must be ≥ 0; ^3P means $S = L = 1$ giving integer J values of 0, 1 or 2 but not $\frac{3}{2}$. Half-integer J's mean odd numbers of electrons and $2S + 1$ even.

4.5 (a) $1s^2 2p$: one open shell electron which has $l = 1$, $s = \frac{1}{2}$ and term $^2P^{\circ}$.

(b) 1s2s3s: 3 open shell electrons, all with $l = 0$ so $L = 0$. All have $s = \frac{1}{2}$, adding them sequentially $S_{12} = 0$ or 1, so $S = S_{12} + s = \frac{1}{2}, \frac{1}{2}, \frac{3}{2}$ (note the two separate values of $S = \frac{1}{2}$). The terms are ^2S, ^2S and ^4S.

(c) 1s2p3p also has 3 open shell electrons. The spin coupling is the same as (b), giving $S = \frac{1}{2}, \frac{1}{2}, \frac{3}{2}$. Coupling $l_1 = 0$, $l_2 = 1$ and $l_3 = 1$ gives $L = 0$, 1 and 2 since l_1 does not contribute. So the resulting terms are ^2S, ^2S, ^4S, ^2P, ^2P, ^4P, ^2D, ^2D and ^4D.

4.6 Only the 3d5p electrons need to be considered as the others are all in closed shells. These have $l_1 = 1$, $s_1 = \frac{1}{2}$ and $l_2 = 2$, $s_2 = \frac{1}{2}$, giving $L = 1, 2, 3$ and $S = 0, 1$. The terms are therefore ^1P, ^1D, ^1F, ^3P, ^3D and ^3F. The lowest is ^3F which has levels $^3F_2$, $^3F_3$ and $^3F_4$.

Chapter 5

5.1 Hα is $n = 2-3$. He II has $Z = 2$ which means the transition will be scaled by a factor Z^2 compared to H I. This means that He II $n = 4-6$ lies close to Hα and that this transition satisfies:

$$\frac{1}{\lambda} = 4R_{He}\left(\frac{1}{n_1^2} - \frac{1}{n_2^2}\right),$$

where $R_{He} = \left(\frac{M_{He}}{M_{He} + m_e}\right)R_\infty$. Assuming $M_{He} = 4M_H$ gives $R_{He} = 109722.4\,\text{cm}^{-1}$ and a wavenumber of $15239.22\,\text{cm}^{-1}$ or $\lambda = 6562.0\,\text{Å}$.

5.2 $1640\,\text{Å}$ corresponds to a He II Balmer α transition with $n = 3-2$. If $n = 3-2$ emissions are observed then Lyman β, $n = 3-1$ at $256\,\text{Å}$ and Lyman α, $n = 2-1$ at $304\,\text{Å}$.

5.3 Shell A: H I and He II recombination spectra.

Shell B: H I and He I recombination spectra.

Shell C: H I recombination spectra only.

Shell D: no recombination spectra will be observed as there are no ions.

5.4 (a) 1s2s has $l_1 = l_2 = L = 0$ and $s_1 = s_2 = \frac{1}{2}$, giving $^1S_0$ and $^3S_1$.

(b) 1s2p has $l_1 = 0$, $l_2 = 1$, so $L = 1$ and $s_1 = s_2 = \frac{1}{2}$, so $S = 0$ or 1. This gives $^1P_1$ and $^3P_0$, $^3P_1$ and $^3P_2$.

Decays in approximate order of strength (strongest first):

$1s2p - 1s^2\ ^1P_1^o - {}^1S_0$, resonance line,

$1s2p - 1s^2\ ^3P_1^o - {}^1S_0$, intercombination line,

$1s2p - 1s^2\ ^3P_2^o - {}^1S_0$, magnetic quadrupole line,

$1s2s - 1s^2\ ^3S_1 - {}^1S_0$, spin-forbidden magnetic transition.

The two $J = 0$ levels cannot decay to the ground state since transitions with $J = 0-0$ do not occur for any selection rules.

5.5 (a)–(b) $1s^22p - 1s2s3s$ is a forbidden transition: parity changes (odd to even) but two electrons jump.

(a)–(c) $1s^22p - 1s2p3p$ is allowed.

(b)–(c) $1s2p3p - 1s2s3s$ is rigorously forbidden by the Laporte rule.

Chapter 6

6.1 (a) Na 3p is $^2P_{\frac{1}{2}}^o$, $^2P_{\frac{3}{2}}^o$; 4d is $^2D_{\frac{3}{2}}$, $^2D_{\frac{5}{2}}$;

Allowed transitions are $^2P_{\frac{1}{2}}^o - {}^2D_{\frac{3}{2}}$, $^2P_{\frac{3}{2}}^o - {}^2D_{\frac{3}{2}}$, $^2P_{\frac{3}{2}}^o - {}^2D_{\frac{5}{2}}$.

(b) Na 3d is $^2D_{\frac{3}{2}}, ^2D_{\frac{5}{2}}$; 5f is $^2F^o_{\frac{5}{2}}, ^2F^o_{\frac{7}{2}}$;

Allowed transitions are $^2D_{\frac{3}{2}} - ^2F^o_{\frac{5}{2}}, ^2D_{\frac{5}{2}} - ^2F^o_{\frac{5}{2}}, ^2D_{\frac{5}{2}} - ^2F^o_{\frac{7}{2}}$.

(c) Na 4s is $^2S_{\frac{1}{2}}$; 4d is $^2D_{\frac{3}{2}}, ^2D_{\frac{5}{2}}$;

Both are even parity so no electric dipole transitions possible. Electric quadrupole transitions have $\Delta J = 0, \pm 1, \pm 2$, parity conserved. Both $^2S_{\frac{1}{2}} - ^2D_{\frac{3}{2}}$ and $^2S_{\frac{1}{2}} - ^2D_{\frac{5}{2}}$ are quadrupole-allowed.

(d) K 4s is $^2S_{\frac{1}{2}}$; 4p is $^2P^o_{\frac{1}{2}}, ^2P^o_{\frac{3}{2}}$;

Allowed transitions are $^2S_{\frac{1}{2}} - ^2P^o_{\frac{1}{2}}, ^2S_{\frac{1}{2}} - ^2P^o_{\frac{3}{2}}$.

All the allowed transitions should be observable in absorption in stellar atmospheres of suitable temperature. Only (d) K 4s – 4p is a possible interstellar absorption feature as all others are between excited states.

6.2* $E(4s) = -35010\,\text{cm}^{-1}$; $E(4p) = -35010 + 12985 = -22025\,\text{cm}^{-1}$; $E(5p) = -35010 + 24701 = -10309\,\text{cm}^{-1}$; $E(6p) = -35010 + 28999 = -6011\,\text{cm}^{-1}$.

The quantum defect, μ_{nl} is given by:

$$E(nl) = -\frac{R_\infty}{(n - \mu_{nl})^2},$$

so $\mu(4p) = 1.768$; $\mu(5p) = 1.737$; $\mu(6p) = 1.727$.

Assuming $\mu(7p) = 1.72$ gives $E(7p) = -3936\,\text{cm}^{-1}$ and 4s–7p at $31074\,\text{cm}^{-1}$. These transitions correspond to the $4\,^2S_{\frac{1}{2}} - n\,^2P^o_{\frac{1}{2}}$ series; for this series the 4s–7p transition is observed at $31070\,\text{cm}^{-1}$.

6.3 Need the splitting $\Delta E = 13080.5 - 12988.7 = 91.8\,\text{cm}^{-1}$.

Level	L	S	J	$\frac{1}{2}[J(J+1) - L(L+1) - S(S+1)]$
$4\,^2P^o_{\frac{3}{2}}$	1	$\frac{1}{2}$	$\frac{3}{2}$	$+\frac{1}{2}$
$4\,^2P^o_{\frac{1}{2}}$	1	$\frac{1}{2}$	$\frac{1}{2}$	-1

So $\frac{3}{2}A' = 91.8\,\text{cm}^{-1}$; $A' = 61.2\,\text{cm}^{-1}$.

6.4 4d has $L = 2$, $S = \frac{1}{2}$ so $J = \frac{3}{2}$ or $\frac{5}{2}$; 3p has $L = 1$, $S = \frac{1}{2}$ so $J = \frac{1}{2}$ or $\frac{3}{2}$. The selection rule is $\Delta J = 0, \pm 1$ so the allowed transitions are $^2D_{\frac{5}{2}} \rightarrow ^2P^o_{\frac{3}{2}}, ^2D_{\frac{3}{2}} \rightarrow ^2P^o_{\frac{3}{2}}$ and $^2D_{\frac{3}{2}} \rightarrow ^2P^o_{\frac{1}{2}}$ (see Fig. 6.5).

If 4d emits to 3p, then the following cascade emissions must also follow: 4p–4s, 4s–3p, 3p–3s and 4p–3s.

The triplet emissions should have different intensities according to their statistical weight. If all three transitions have the same intensity, they are optically thick and column density of Na cannot be determined from this observation.

6.5*

$$E(nl) = -\frac{R_\infty Z_{eff}^2}{(n - \mu_{nl})^2},$$

where $Z_{eff} = 4$. The quantum defect gives the departure of the energy levels of the outer electron from the pure hydrogenic case. It is largely determined by penetration effect, with the values generally decreasing with l; for a given l it usually only depends weakly on n.

Using $E = 520178\,\text{cm}^{-1}$ and $n = 2$ gives $\mu(2s) = 0.163$; $E = 217329\,\text{cm}^{-1}$ and $n = 3$ gives $\mu(3s) = 0.158$. So using $\mu(4s) = 0.155$ and $n = 4$ gives $E(4s) = 118763\,\text{cm}^{-1}$. Observed value is $118830\,\text{cm}^{-1}$.

6.6 (a) Hund's rules give $^4S^o$ lowest, then $^2D^o$ and $^2P^o$ highest.

(b) Levels are $^4S^o_{\frac{3}{2}}$, $^2D^o_{\frac{3}{2}}$, $^2D^o_{\frac{5}{2}}$, $^2F^o_{\frac{5}{2}}$ and $^2F^o_{\frac{7}{2}}$.

Level	L	S	J	$\frac{1}{2}[J(J+1) - L(L+1) - S(S+1)]$
$^2D^o_{\frac{3}{2}}$	2	$\frac{1}{2}$	$\frac{3}{2}$	-1.5
$^2D^o_{\frac{5}{2}}$	2	$\frac{1}{2}$	$\frac{5}{2}$	$+1$
$^2F^o_{\frac{5}{2}}$	3	$\frac{1}{2}$	$\frac{5}{2}$	-2
$^2F^o_{\frac{7}{2}}$	3	$\frac{1}{2}$	$\frac{7}{2}$	$+1.5$

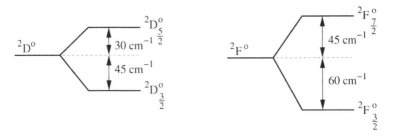

6.7 $^3F^o_2$ means $S = 1$, $L = 3$, $J = 2$ and odd parity. $^3F^o$ term also gives $^3F^o_3$ and $^3F^o_4$. The term could arise from the configuration $1s^22s^22p3d$ (other answers are possible).

In order of diminishing strength:

$^3F^o_2 - {}^3D_3$ is dipole-allowed and will be strong.

$^3F^o_2 - {}^1D_3$ has $\Delta S \neq 0$ and is an intercombination line.

$^3F^o_2 - {}^3P_2$ has $\Delta L = 2$ and is a forbidden transition.

$^3F^o_2 - {}^3P^o_2$ does not change parity so is completely dipole-forbidden by the Laporte rule.

Chapter 7

7.1 (a) $4s^2$ gives $^1S_0$; $4s4p$ gives $^1P_1^o$, $^3P_0^o$, $^3P_1^o$ and $^3P_2^o$. Decays in approximate order of strength (strongest first):

$4s4p - 4s^2$ $^1P_1^o - {}^1S_0$, resonance line,
$4s4p - 4s^2$ $^3P_1^o - {}^1S_0$, intercombination line,
$4s4p - 4s^2$ $^3P_2^o - {}^1S_0$, magnetic quadrupole line.
$^3P_0^o$ does not have an allowed decay to the $4s^2$ $^1S_0$ ground state.

(b) $4s4f$ gives $^1F_3^o$, $^3F_2^o$, $^3F_3^o$ and $^3F_4^o$; $4s5d$ gives $^1D_2$, $^3D_1$, $^3D_2$ and $^3D_3$. Allowed dipole transitions:

$^1F_3^o - {}^1D_2$, $^3F_2^o - {}^3D_1$, $^3F_2^o - {}^3D_2$, $^3F_2^o - {}^3D_3$, $^3F_3^o - {}^3D_2$, $^3F_3^o - {}^3D_3$, $^3F_4^o - {}^3D_3$.

All the allowed transitions could be observed in emission as part of the Ca I recombination spectra in H II regions. The transitions (a) could be seen in absorption from the ISM.

7.2 N^+ is C-like: $1s^2 2s^2 2p^2$. Gives terms ^3P, ^1D and ^1S as C. Hund's rule (1) means ^3P will be the ground state.
Transitions $1s^2 2s^2 2p^2(^3P)5f \rightarrow 4d$. Need to consider coupling between ^3P core and the extra electron.

Upper state: ^3P gives $L = 1$, $S = 1$; $5f$ gives $l = 3$, $s = \frac{1}{2}$. So $L = 2, 3, 4$; $S = \frac{1}{2}, \frac{3}{2}$ giving $^2D^o$, $^4D^o$, $^2F^o$, $^4F^o$, $^2G^o$ and $^4G^o$.

Lower state: ^3P gives $L = 1$, $S = 1$; $4d$ gives $l = 2$, $s = \frac{1}{2}$. So $L = 1, 2, 3$; $S = \frac{1}{2}, \frac{3}{2}$ giving ^2P, ^4P, ^2D, ^4D, ^2F and ^4F.

Allowed transitions, 12 in total:

$^2D^o \rightarrow {}^2P$, $^2D^o \rightarrow {}^2D$, $^2D^o \rightarrow {}^2F$, $^2F^o \rightarrow {}^2D$, $^2F^o \rightarrow {}^2F$, $^2G^o \rightarrow {}^2F$, $^4D^o \rightarrow {}^4P$, $^4D^o \rightarrow {}^4D$, $^4D^o \rightarrow {}^4F$, $^4F^o \rightarrow {}^4D$, $^4F^o \rightarrow {}^4F$, $^4G^o \rightarrow {}^4F$.

(In fact inclusion of fine structure effects leads to 71 distinct transitions.) The strongest transition is $^4G^o \rightarrow {}^4F$ as:

(a) $^4G^o$ has the highest statistical weight and therefore occupancy;
(b) $^4G^o$ only emits to ^4F. Terms D^o and F^o have competing decay routes.

7.3 The transitions lead to the following energies relative to the $2p^2$ $^3P_0$ ground level:

Level	Energy (eV)	g	Population, $T = 10000\,\text{K}$
$2p^2\,{}^3P_0$	0.000	1	1.000
$2p^2\,{}^3P_1$	0.014	3	2.958
$2p^2\,{}^3P_2$	0.038	5	4.814
$2p^2\,{}^1D_2$	2.514	5	0.405
$2p^2\,{}^1S_0$	5.356	1	0.0047
$2p3p\,{}^3P_2$	40.849	5	9.1×10^{-18}

Note that several of these levels can be calculated in different ways which provides a consistency and accuracy check. The level ${}^{2S+1}L_J$ has statistical weight $g = 2J + 1$; individual degeneracies are given in the table. Population relative to the ground state is given by Boltzmann's distribution $g \exp(-\frac{E}{kT})$, where it has been assumed that for the ground state $E = 0$ and $g = 1$. Results for $T = 10000\,\text{K}$ are given in the table. At a typical nebular temperature of $10000\,\text{K}$ the fine structure levels of ${}^3P$ are populated almost in proportion to their statistical weights, so the populations of these levels are not a good indicator of temperature. The ${}^1D_2$ and ${}^1S_0$ levels have reasonable thermal populations, which are temperature sensitive, so these levels will give reasonably strong transitions which can be used to determine temperature.

The thermal population of $2p3p\,{}^3P_0$ is negligible; this level can only be populated by some other mechanism.

Chapter 8

8.1

$$E_n = -R_\infty \frac{Z^2}{n^2}.$$

O VIII Lyα is at $4032846\,\text{cm}^{-1}$, which corresponds to $500\,\text{eV}$. This lies in the X-ray and such transitions can only be observed using satellites.

8.2 $1s^2 2s^2 2p^6 3s^2 3p^6 3d^{10} 4s^2 4p^6 4d^{10} 4f^{14} 5s^2 5p^6 5d^9 6s^2$. Hg$^+$ $5d^{10}6s$ gives term ${}^2S$ and level ${}^2S_{\frac{1}{2}}$. Both this configuration and the $5d^9 6s^2$ configuration are even parity so dipole transitions between them are forbidden by the Laporte rule. The $5d^9 6s^2$ ${}^2D_{\frac{5}{2}}$ state is metastable since it cannot decay by a dipole transition.

Chapter 9

9.1 Boltzmann's distribution gives the probability, P_J, of being in level J

$$\frac{P_J}{P_0} = (2J + 1)\exp\left(-\frac{BJ(J+1)}{kT}\right).$$

For a maximum:

$$\frac{d\frac{P_J}{P_0}}{dJ} = 2\exp\left[-\frac{BJ(J+1)}{kT}\right] - (2J+1)^2\frac{B}{kT}\exp\left[-\frac{BJ(J+1)}{kT}\right] = 0.$$

$$(2J+1)^2 = \frac{2kT}{B}, \quad J = \left(\frac{kT}{2B}\right)^{\frac{1}{2}} - \frac{1}{2}.$$

9.2 Using the formula derived in answer 9.1:
(a) $J = 1$; (b) $J = 7$; (c) $J = 22$ or 23.
Using Eq. (9.34):
(a) Effectively zero; (b) 0.003%; (c) 35%.

9.3 Zero point energy $= \frac{\hbar\omega}{2} = 1037.8\,\text{cm}^{-1}$. Within the harmonic approximation $\omega = \left(\frac{k}{\mu}\right)^{\frac{1}{2}}$, so:

$$\omega_{13} = \omega_{12}\left(\frac{\mu_{12}}{\mu_{13}}\right)^{\frac{1}{2}} = 2075.5\left(\frac{12}{12+16} \cdot \frac{13+16}{13}\right)^{\frac{1}{2}} = 2069.4\,\text{cm}^{-1}.$$

Zero point energy is $= \frac{\hbar\omega}{2} = 1014.7\,\text{cm}^{-1}$. This assumes that k is the same for both $^{12}\text{CH}^+$ and $^{13}\text{CH}^+$.

$^{13}\text{CH}^+$ may be optically thin when $^{12}\text{CH}^+$ is optically thick.

At low temperature, fractionation effects will increase the ratio $n(^{13}\text{CH}^+)$ to $n(^{12}\text{CH}^+)$ compared to $n(^{13}\text{C})$ to $n(^{12}\text{C})$ since $^{13}\text{CH}^+$ has the lower zero point energy.

Chapter 10

10.1 The gaps between the lines are 3.83, 7.70, 3.87 and 3.91 cm^{-1}, which correspond (approximately) to $2B, 4B, 2B, 2B$. The $4B$ gap is the band centre and the transitions are therefore P(2), P(1), R(0), R(1) and R(2), respectively. CO will absorb in the infrared both in the ISM and in the atmosphere of cool stars. CO infrared emissions are less common but can occur in the warm ISM (e.g. shocked regions or nebulae).

10.2 (a) Assuming a rigid rotor, the $J = 1-0$ transition at $2B$ and $B \propto \mu^{-1}$.

(b) Assuming a harmonic oscillator, the $v = 1-0$ transition is at $\hbar\omega$ and $\omega \propto \mu^{-\frac{1}{2}}$.

$$\mu_{12} = \frac{M_{12}M_O}{M_{12} + M_O}; \quad \mu_{13} = \frac{M_{13}M_O}{M_{13} + M_O};$$

$$\frac{\mu_{12}}{\mu_{13}} = \frac{M_{12}(M_{13} + M_O)}{M_{13}(M_{12} + M_O)} = 0.9560,$$

assuming $M_{12} = 12$, $M_{13} = 13$ and $M_O = 16$ u. $J = 1-0$ for $^{13}C^{16}O$ will be at $3.86 \times 0.9560 = 3.69\,cm^{-1}$.

$\hbar\omega$ for $^{13}C^{16}O$ will be at $2170 \times \sqrt{0.9560} = 2122\,cm^{-1}$.

The abundance of ^{13}C is much lower than that of ^{12}C so observable lines for $^{13}C^{16}O$ may well be optically thick for $^{12}C^{16}O$.

10.3 Rigid rotor transitions at $2BJ'$ so $J = 5-4$ will be at $10B = 19.3\,cm^{-1}$ and $J = 20-19$ will be at $40B = 77.2\,cm^{-1}$. The estimate for $J = 5-4$ will be more accurate since the contribution from centrifugal distortion will be lower.

10.4 (a) Far-infrared, rotational spectrum:

In emission: $J = 1-0$ at $2B = 23.88\,cm^{-1}$;
In absorption: $J = 2-1$ at $4B = 47.76\,cm^{-1}$.

(b) Mid-infrared, vibration–rotation transitions, observed only in absorption:

P(1) at $\omega - 2B = 2051.6\,cm^{-1}$; R(1) at $\omega + 4B = 2123.3\,cm^{-1}$.

10.5* Even for $J \simeq 30$, CO behaves as a near rigid rotor with transitions spaced by $2B_0 \simeq 3.8\,cm^{-1}$. Measuring approximate wavelengths from Fig. 7.6 gives:

$\lambda\,(\mu m)$	$\hbar\omega\,(cm^{-1})$	$J' - J''$
186	53.8	14–13
174	57.5	15–14
162	61.7	16–15
153	65.4	17–16
145	68.9	18–17
138	72.5	19–18
131	76.3	20–19
123	81.3	21–20
118	84.7	22–21
113	88.5	23–22
108	92.6	24–23

10.6* Figure 10.9: central feature (where a line is missing) at $\lambda_0 \simeq 4.66$. The R-branch lies at $\lambda < \lambda_0$ with lines up to the R(0), R(1), R(2), R(3), R(4), R(5) up to the maximum at R(6). The P-branch lies at $\lambda > \lambda_0$ with lines P(1), P(2), P(3), P(4) up to the maximum at P(5). R(6) means $J' - J'' = 7 - 6$ i.e. $J'' = 6$; P(5) means $J' - J'' = 4 - 5$ so $J'' = 5$. So the highest occupied level in the absorbing gas is $J'' = 5$ or 6. Using the formula derived in answer 9.1 gives $T \approx 168$ K for $J_{max} = 5$ or 235 K for $J_{max} = 6$.

Note: this is very approximate. A reliable temperature estimate can be obtained by a fit to the intensity of all the transitions in the band.

10.7 Denote the energy levels $E_X(v'', J'')$ and $E_B(v', J')$ for the ground and excited electronic states only. As the H_2 is cold, only $J'' = 0$ or 1 will be occupied. Then, using constants for the appropriate electronic state taken from Table 10.2,

$$E_X(0,0) = \frac{w_e''}{2} = 2200.6 \, \text{cm}^{-1},$$
$$E_X(0,1) = E_X(0,0) + 2B_e'' = 2322.3 \, \text{cm}^{-1},$$
$$E_C(0,0) = T_e + \frac{w_e''}{2} = 101311.7 \, \text{cm}^{-1},$$
$$E_C(0,1) = E_C(0,0) + 2B_e' = 101374.4 \, \text{cm}^{-1},$$
$$E_C(0,2) = E_C(0,0) + 6B_e' = 101499.9 \, \text{cm}^{-1},$$
$$E_C(1,0) = E_C(0,0) + w_e' = 103755.5 \, \text{cm}^{-1},$$
$$E_C(1,1) = E_C(0,1) + w_e' = 103818.2 \, \text{cm}^{-1},$$
$$E_C(1,2) = E_C(0,2) + w_e' = 103943.6 \, \text{cm}^{-1},$$
$$E_C(2,0) = E_C(1,0) + w_e' = 106199.2 \, \text{cm}^{-1},$$
$$E_C(2,1) = E_C(1,1) + w_e' = 106261.9 \, \text{cm}^{-1},$$
$$E_C(2,2) = E_C(1,2) + w_e' = 106387.4 \, \text{cm}^{-1}.$$

The conditions specified give twelve transitions. These occur at:

$$(0,0) \, Q(1) \text{ at } \omega = E_C(0,1) - E_X(0,1) = 99052.1 \, \text{cm}^{-1},$$
$$(0,0) \, R(0) \text{ at } \omega = E_C(0,1) - E_X(0,0) = 99173.8 \, \text{cm}^{-1},$$
$$(0,0) \, P(1) \text{ at } \omega = E_C(0,0) - E_X(0,1) = 98989.4 \, \text{cm}^{-1},$$
$$(0,0) \, R(1) \text{ at } \omega = E_C(0,2) - E_X(0,1) = 99177.6 \, \text{cm}^{-1},$$
$$(1,0) \, Q(1) \text{ at } \omega = E_C(1,1) - E_X(0,1) = 101495.9 \, \text{cm}^{-1},$$
$$(1,0) \, R(0) \text{ at } \omega = E_C(1,1) - E_X(0,0) = 101617.6 \, \text{cm}^{-1},$$
$$(1,0) \, P(1) \text{ at } \omega = E_C(1,0) - E_X(0,1) = 101433.2 \, \text{cm}^{-1},$$
$$(1,0) \, R(1) \text{ at } \omega = E_C(1,2) - E_X(0,1) = 101621.3 \, \text{cm}^{-1},$$
$$(2,0) \, Q(1) \text{ at } \omega = E_C(2,1) - E_X(0,1) = 103939.6 \, \text{cm}^{-1},$$
$$(2,0) \, R(0) \text{ at } \omega = E_C(2,1) - E_X(0,0) = 104061.3 \, \text{cm}^{-1},$$
$$(2,0) \, P(1) \text{ at } \omega = E_C(2,0) - E_X(0,1) = 103876.9 \, \text{cm}^{-1},$$
$$(2,0) \, R(1) \text{ at } \omega = E_C(2,2) - E_X(0,1) = 104065.1 \, \text{cm}^{-1}.$$

INDEX

Printed in the United States
By Bookmasters